Statistical Tutor for Johnson and Kuby's

JUST THE ESSENTIALS

—————————— of ——————————

E l e m e n t a r y   S t a t i s t i c s

S E C O N D   E D I T I O N

D1495411

P A T R I C I A   K U B Y

DUXBURY PRESS

An Imprint of Brooks/Cole Publishing Company
I(T)P® An International Thomson Publishing Company

Pacific Grove • Albany • Belmont • Bonn • Boston • Cincinnati • Detroit
Johannesburg • London • Madrid • Melbourne • Mexico City • New York
Paris • Singapore • Tokyo • Toronto • Washington

Assistant Editor: *Cindy Mazow*
Editorial Assistant: *Rita Jaramillo*
Marketing: *Marcy Perman*
Production Editor: *Mary Vezilich*
Printing and Binding: *Webcom Limited*

*For more information, contact Duxbury Press at Brooks/Cole Publishing Company,
or electronically at* **http://www.duxbury.com**

BROOKS/COLE PUBLISHING COMPANY
511 Forest Lodge Road
Pacific Grove, CA 93950
USA

International Thomson Editores
Seneca 53
Col. Polanco
11560 México, D. F., México

International Thomson Publishing Europe
Berkshire House 168-173
High Holborn
London WC1V 7AA
England

International Thomson Publishing Japan
Hirakawacho Kyowa Building, 3F
2-2-1 Hirakawacho
Chiyoda-ku, Tokyo 102
Japan

Thomas Nelson Australia
102 Dodds Street
South Melbourne, 3205
Victoria, Australia

International Thomson Publishing Asia
60 Albert Street
#15-01 Albert Complex
Singapore 189969

Nelson Canada
1120 Birchmount Road
Scarborough, Ontario
Canada M1K 5G4

International Thomson Publishing GmbH
Königswinterer Strasse 418
53227 Bonn
Germany

Printed in Canada

5   4   3   2   1

ISBN 0-534-36165-X

CONTENTS

Preface   v

Solutions and Student Annotations

Chapter 1   Statistics   2
Chapter 2   Descriptive Analysis and Presentation
              of Single-Variable Data   8
Chapter 3   Descriptive Analysis and Presentation
              of Bivariate Data   51
Chapter 4   Probability   79
Chapter 5   Probability Distribution
              (Discrete Variables)   104
Chapter 6   Normal Probability Distributions   133
Chapter 7   Sample Variability   155
Chapter 8   Introduction to Statistical
              Inferences   171
Chapter 9   Inferences Involving One Population   205
Chapter 10   Inferences Involving Two Populations   237
Chapter 11   Applications of Chi-Square   272

Introductory Concepts
              Summation Notation   294
              Using the Random Number Table   302
              Round-Off Procedure   305

Review Lessons
              The Coordinate-Axis System and the
              Equation of a Straight Line   308
              Tree Diagrams   316
              Venn Diagrams   321
              The Use of Factorial Notation   324

Answers to Introductory Concepts and Review Lessons Exercises
              Summation Notation   326
              Using the Random Number Table   326
              Round-Off Procedure   327
              The Coordinate-Axis System and the
              Equation of a Straight Line   327
              Tree Diagrams   331
              Venn Diagrams   334
              The Use of Factorial Notation   336

PREFACE

This <u>Statistical Tutor</u> contains solutions for all of the margin and odd-numbered exercises in <u>Just the Essentials of Elementary Statistics</u>, 2nd edition, as well as student information and assistance. Included at the end of the manual are sections covering Introductory Concepts and Review Lessons on various algebraic and/or basic statistical concepts.
For each chapter, the following are provided:
    1) Chapter Preview
    2) Chapter Solutions for Margin and Odd-Numbered Exercises
    3) Student Annotations.

Student annotations are printed inside a border and placed before related exercises.
There are several ways for a student to use this manual. One possibility that has proved to be beneficial contains the following steps. Begin by reading the exercise under consideration in the text and attempting to answer the question. Locate the exercise number in the manual. If the exercise was answered, compare solutions. If solutions are not similar or difficulty has been encountered, read the boxed annotations before the exercise in detail.
Try the exercise again. If difficulty remains at this point, review the boxed annotations and the solution. Ask the instructor for aid or verification to get needed additional information.
In several exercises, a concluding question is asked to determine if the overall concept is comprehended. These questions are marked with an asterisk *. Locations of the corresponding answers are given and marked with an * and the exercise number.

CHAPTER 1 ▽ STATISTICS

Chapter Preview

The purpose of Chapter 1 is to present:
1. an initial image of statistics,
2. its basic vocabulary and definitions,
3. basic ideas and concerns about the processes used to obtain sample data.

SECTION 1.1 MARGIN EXERCISES

The articles in exercises 1.1 & 1.2, give information about the sample (the number of people surveyed). Be watchful of articles that do not give any of this information. Sometimes not knowing something about the sample or survey size causes a question of credibility.

1.1 a. King-Size Company customers
b. 10,000
c. 99% of the 10,000 customers considered airline seating cramped.
d. More than one answer was allowed.

1.2 a. American men
b. Each of the samples described in the article were selected from part of the population, leaving a large proportion of the population with no representation.

SECTION 1.1 EXERCISES

Descriptive Statistics - refers to the techniques and methods for organizing and summarizing the information obtained from the sample.

Inferential Statistics - refers to the techniques of interpreting and generalizing about the population based on the information obtained from the sample.

1.3   a. inferential   b. descriptive

1.5   a. Insurance companies have a major problem.

   b. $\dfrac{58 - 39}{39} = \dfrac{19}{39} = 48.7\%$ increase

   c. $\dfrac{593 - 490}{490} = \dfrac{103}{490} = 21\%$ increase

   d. 48.7% vs. 21%
   e. No, percentages alone do not indicate size of the numbers.

1.7   Measure the seat dimensions on the various airplanes and compare them to the dimensions that passengers do find comfortable.

SECTION 1.2 MARGIN EXERCISES

1.8   Possibilities: marital status, ZIP code, gender, highest level of education

1.9   Possibilities: annual income, age, distance to store, amount spent

1.10 a. 1) Did most of your family eat dinner together last night?
   2) How important to you is eating dinner with your family?
   3) In the last seven days, how many evenings did most of your family eat dinner together?
   4) How long would you say dinner usually lasts when you eat together?

   b. attribute, attribute, numerical, numerical

SECTION 1.2 EXERCISES

<u>Population</u> - the collection of all individuals, objects, or scores whose properties are under consideration.

<u>Parameter</u> - a number calculated from the population of values.

<u>Sample</u> - that part of the population from which the data values or information is obtained.

<u>Statistic</u> - a number calculated from the sample values.

NOTE: <u>Parameters</u> are calculated from <u>populations</u>; both begin with p. <u>Statistics</u> are calculated from <u>samples</u>; both begin with s.

1.11 a. All individuals who have hypertension and use prescription drugs to control it (a very large group)
   b. The 5,000 people in the study
   c. The proportion of the population for which the drug is effective
   d. The proportion of the sample for which the drug is effective, 80%
   e. No, but it is estimated to be approximately 80%

A <u>variable</u> is the characteristic of interest (ex. height), where <u>data</u> is a value for the variable (ex. 5'5"). A variable varies (heights vary), that is, heights can take on different values. Data (singular) such as 5'5" (one person's height) is constant; it does not change in value for a specific subject.

...

A Statistic and Its Corresponding Variable

A statistic is a single numerical characteristic (number) where its corresponding variable is the overall characteristic of interest (words).

Suppose that the average age of New York State college students is 24. 24 is a statistic and age of New York State college students is its corresponding variable.

An attribute variable can take on any qualitative or "numerical" qualitative information (ex. kinds of fruit, types of music, religious preference, model year - most answers are in words, although model year would have "numerical" answers such as "1990").

A numerical variable can take on any quantitative information. This includes any count-type and measurable-type data (ex. number of children in a family, place value in a race, race time, age, height).

1.13 a. All assembled parts from the assembly line
     b. infinite   c. the parts checked
     d. attribute, attribute (it identifies the assembler), quantitative

1.15 a. numerical   b. attribute   c. numerical
     d. attribute   e. numerical   f. numerical

1.17 a. The population contains all objects of interest, while the sample contains only those actually studied.
     b. convenience, availability, practicality

SECTION 1.3 EXERCISES

1.19 Group 2, the football players, because their weights cover a wider range of values, probably 175 to 300+, while the cheerleaders probably all weigh between 110 and 150.

1.21 A lack of variability would indicate all students attained very similar scores, even when they do not all know the same amount.

SECTION 1.4 MARGIN EXERCISES

1.23 a. All adults who deal with stockbrokers
   b. Gender, answers to questions dealing with risk tolerance, information about explanations regarding investments and bonds, number of interruptions

SECTION 1.4 EXERCISES

1.25 a. There are three choices for the first pick and then still three choices for the second pick; therefore 3 times 3 gives 9 different samples.

   b. (1,1), (1,2), (1,3), (1,4),
      (2,1), (2,2), (2,3), (2,4),
      (3,1), (3,2), (3,3), (3,4),
      (4,1), (4,2), (4,3), (4,4)

   c. (1,1,1), (1,1,2), (1,1,3),
      (1,2,1), (1,2,2), (1,2,3),
      (1,3,1), (1,3,2), (1,3,3),

      (2,1,1), (2,1,2), (2,1,3),
      (2,2,1), (2,2,2), (2,2,3),
      (2,3,1), (2,3,2), (2,3,3),

      (3,1,1), (3,1,2), (3,1,3),
      (3,2,1), (3,2,2), (3,2,3),
      (3,3,1), (3,3,2), (3,3,3)

d. (1,1,1), (1,1,2), (1,1,3), (1,1,4),
   (1,2,1), (1,2,2), (1,2,3), (1,2,4),
   (1,3,1), (1,3,2), (1,3,3), (1,3,4),
   (1,4,1), (1,4,2), (1,4,3), (1,4,4),

   (2,1,1), (2,1,2), (2,1,3), (2,1,4),
   (2,2,1), (2,2,2), (2,2,3), (2,2,4),
   (2,3,1), (2,3,2), (2,3,3), (2,3,4),
   (2,4,1), (2,4,2), (2,4,3), (2,4,4),

   (3,1,1), (3,1,2), (3,1,3), (3,1,4),
   (3,2,1), (3,2,2), (3,2,3), (3,2,4),
   (3,3,1), (3,3,2), (3,3,3), (3,3,4),
   (3,4,1), (3,4,2), (3,4,3), (3,4,4),

   (4,1,1), (4,1,2), (4,1,3), (4,1,4),
   (4,2,1), (4,2,2), (4,2,3), (4,2,4),
   (4,3,1), (4,3,2), (4,3,3), (4,3,4),
   (4,4,1), (4,4,2), (4,4,3), (4,4,4)

1.27 A simple random sample would be very difficult to obtain from an extremely large or spread out population.

1.29 a. judgment sample
   b. No, statistical inference requires that the sample design be a probability sample.

1.31 Only people with telephones and listed phone numbers will be considered.

SECTION 1.5 MARGIN EXERCISES

1.33 a. probability   b. statistics

SECTION 1.5 EXERCISES

> Statistics - allows you to make inferences or generalizations about the population based on a sample.
>
> Probability - the chance that something will happen when you know **all** the possibilities (you know all the possibilities when you have the population).

1.35 a. statistics   b. statistics
     c. probability   d. statistics

SECTION 1.6 EXERCISES

1.37 Draw graphs, print charts, calculate statistics

CHAPTER EXERCISES

1.39 a. The proportion of registered voters who will vote for the candidate
     b. 23% (35/150)
     c. No. If the sample is representative of the population, the candidate will not win.

1.41 a. C is a piece of data.
     b. What percent of the mothers in the sample complimented their child one or more times yesterday?
     c. What percent of all mothers complimented their child one or more times yesterday?

1.43 qualitative

1.45 a. Numerical data, since the number of women who avoided rape was counted for two different groups, one group which resisted and the other group which did not resist.
     b. Judgment sample.

1.47 Each will have different examples.

CHAPTER 2 ▽ DESCRIPTIVE ANALYSIS AND PRESENTATION OF SINGLE-VARIABLE DATA

Chapter Preview

Chapter 2 deals with the presentation of data that were obtained through the various sampling techniques discussed in Chapter 1. The four major areas for presentation and summary of the data are:
1. graphical displays,
2. measures of central tendency,
3. measures of variation, and
4. measures of position.

SECTION 2.1 MARGIN EXERCISES

2.1 HOW US CONSUMERS PAID FOR GOODS AND SERVICES IN 1993

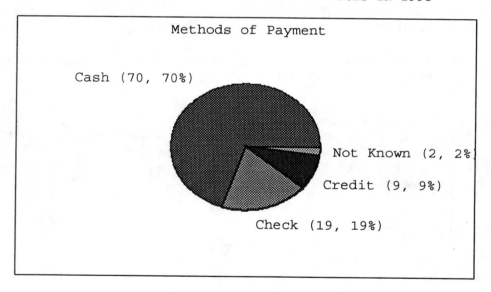

2.2 HOW US CONSUMERS PAID FOR GOODS AND SERVICES IN 1993

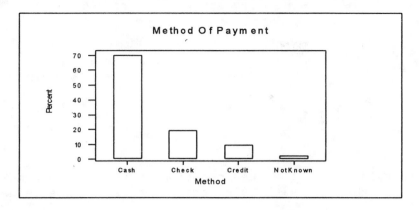

2.3 The circle graph makes it easier to visually compare the relative sizes of the parts to each other and the size of each part to the whole.

2.4

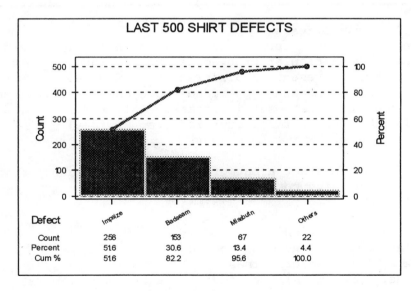

2.5 **Points Scored per Game by Basketball Team**

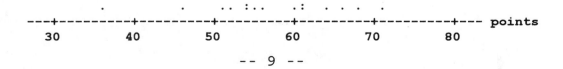

2.6 96: leaf of 6 is placed on the 9 stem
66: leaf of 6 is placed on the 6 stem

2.7 Points scored per game
```
3 | 6
4 | 6
5 | 6 4 5 4 2 1
6 | 1 1 8 0 6 1 4
7 | 1
```


2.8 Each leaf value is in the ones position of the complete data value. For example, with 9|8, the 8 represents the value 8, as in 98.

SECTION 2.1 EXERCISES

MINITAB - Statistical software
Data is entered by use of a spreadsheet divided into columns and rows. Data for each particular problem is entered into its own column. Each column represents a different set of data. Be sure to name the columns in the space provided above the first row, so that you know where each data set is located. All MINITAB commands may be abbreviated to their first four letters.
                    (; = sub command, . = end)

MINITAB commands to construct a bar graph with the frequencies in C1 and their corresponding categories in C2:
      CHARt C1 * C2;
      BAR;
      TITLe 'title of your choice'.
For other variations, use the Menu commands:
GRID(under FRAME) - to add gridlines
DATA LABELS(under ANNOTATION) - to add counts to tops of bars

2.9 a.

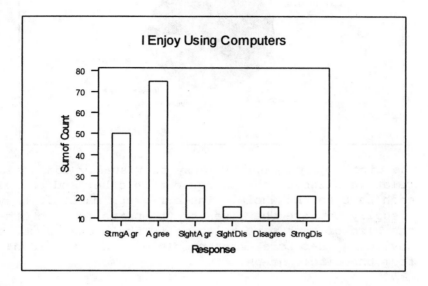

b. 25%, 37.5%, 12.5%, 7.5%, 7.5%, 10%

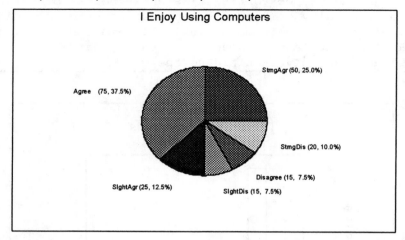

I Enjoy Using Computers

StrngAgr (50, 25.0%)

Agree (75, 37.5%)

StrngDis (20, 10.0%)

Disagree (15, 7.5%)

SlghtAgr (25, 12.5%)

SlghtDis (15, 7.5%)

c. The circle graph makes it easy to visually compare the
relative sizes of the parts to each other and the size of
each part to the whole. The bar graph makes it easy to
visually compare the sizes of the parts to each other, but
the size of each part relative to the whole is not as
obvious. Therefore, most people are able to "read" more
from the circle graph.

Frequency = (total #)(percentage)

2.11 a. The frequencies are 54, 138, 120, 276 and 12.

b.

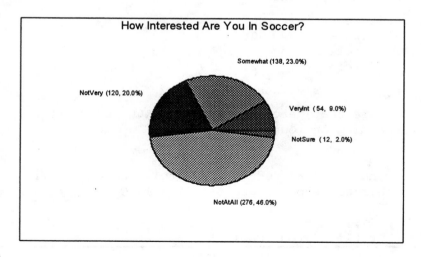

How Interested Are You In Soccer?

Somewhat (138, 23.0%)

NotVery (120, 20.0%)

VeryInt (54, 9.0%)

NotSure (12, 2.0%)

NotAtAll (276, 46.0%)

c.

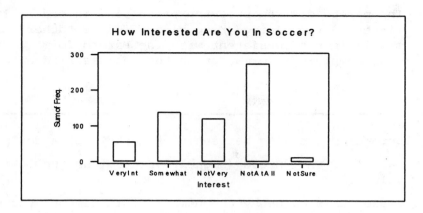

MINITAB commands to construct a Pareto diagram can be found in JES2*-p32.
*(JES2 denotes the textbook Just the Essentials of Elementary Statistics, 2nd edition)

The Pareto command generates bars, starting with the largest category, until the cumulative percent surpasses 95. The remaining categories are grouped together into a bar named "Others".
NOTE: Pareto diagrams are primarily used for quality control applications and therefore MINITAB's PARETO command identifies the categories as "Defects", even when they may not be defects.

2.13 a.

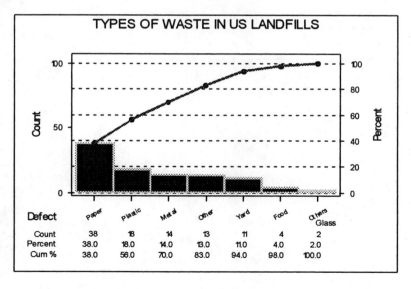

Defect	Paper	Plastic	Metal	Other	Yard	Food	Others Glass
Count	38	18	14	13	11	4	2
Percent	38.0	18.0	14.0	13.0	11.0	4.0	2.0
Cum %	38.0	56.0	70.0	83.0	94.0	98.0	100.0

b. The pareto diagram puts the categories in order, largest first. A very large group classified as "other" in the middle is not consistent with the Pareto format. We need more complete information on the unclassified data.

Picking increments (spacing between tick marks) for a dot plot

1. Calculate the spread (highest value minus the lowest value).

2. Divide this value by the number of increments you wish to show (no more than 7 usually).

3. Use this increment size or adjust to the nearest number that is easy to work with (5, 10, etc.).

OR:

MINITAB commands to construct a dotplot can be found in JES2-p33. Another variation:

```
DOTPlot C1;    -   You specify the increments (spacing
INCRement 1.       between the tick marks), for the dot plot.
```

2.15 Street speed of cars

```
                 .  .
               .  .  .
             . . . .  . . .
        .  . . . . . . . . . .  .      .      .            .
      ---+----+----+----+----+----+----+----+---
        20   25   30   35   40   45   50
            Speed (miles per hour)
```

2.17 a. 15
b. 11.2, 11.2, 11.3, 11.4, 11.7
c. 15.6
d. 13.7; 3

```
STEM-AND-LEAF DISPLAYS
    1. Find the lowest and highest data values.

    2. Decide in what "place value" position, the data values
       will be split.

    3. Stem = leading digit(s)

    4. Leaf = trailing digit(s) (if necessary, data is first
       "rounded" to the desired position)

    5. Sort stems and list.

    6. Split data values accordingly, listing leaves on the
       appropriate stem.
OR:

MINITAB commands to construct a stem-and-leaf diagram can be found
in JES2-p36.
```

2.19 One-way travel time

```
0 |5 5
1 |5 5 5 0
2 |0 0 5 0 5 5 5 5 5 0 5 0 0 0 0 5 0 0 0
3 |0 5 0
4 |0 5
```

```
The column on the left of the stem-and-leaf display is the
cumulative count of the data from the top (low-value) down and the
bottom (high-value) up until the median class is reached.  The
number of data values for the median class is in parentheses.
```

2.21 a. The spacing between the stems is 0.1; i.e., 59 and 60 are
            5.9 and 6.0.
         b. The place value of the leaves is in the hundredths place;
            i.e., 59|7 is 5.97.
         c. 16
         d. 5.97, 6.01, 6.04, 6.08

2.23

x	f
0	2
1	5
2	3
3	0
4	2
	12

2.24 a. f is frequency, therefore values of 70 or more but less than 80 occurred 8 times.

b. $\Sigma f = 1 + 3 + 8 + 5 + 2 = 19$
Σf = sum of all the frequencies = sample size, or number of data

c. 19

2.25 a. $5 \leq x < 8$

b.
class interval	frequency	
$0 \leq x < 2$	30	(500·0.06)
$2 \leq x < 5$	35	(500·0.07)
$5 \leq x < 8$	75	(500·0.15)
$8 \leq x < 11$	60	(500·0.12)
$11 \leq x$	300	(500·0.60)
	500	

2.26 a. Class #4; $65 \leq x < 75$

b. This class contains all values greater than or equal to 65 and also less than 75(does not include 75).

c. Difference between upper and lower class boundaries.

i. Subtracting the lower class boundary from the upper class boundary for any one class

ii. Subtracting a lower class boundary from the next consecutive lower class boundary

iii. Subtracting an upper class boundary from the next consecutive upper class boundary

iv. Subtracting a class mark from the next consecutive class mark.

2.27

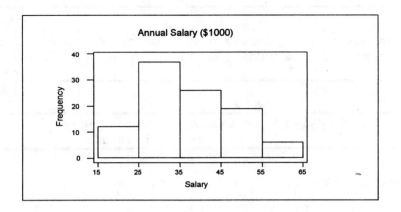

2.28 The shapes are the same but Figure 2-10 uses class marks on the horizontal scale, whereas Figure 2-11 uses the class boundaries. Figure 2-10 uses frequency(count) on the vertical axis, whereas Figure 2-11 has turned those counts into percents.

2.29 Symmetric: weight of dry cereal per box, breaking strength of certain type of string
Uniform: tossing a die several hundred times
Skewed Right: salaries, high school class sizes
Skewed left: hour exam scores
Bimodal: heights, weights for groups containing both male and female
J-shaped: amount of television watched per day

SECTION 2.2 EXERCISES

Frequency distributions can be either grouped or ungrouped. Ungrouped frequency distributions have single data values as *x* values. Grouped frequency distributions have intervals of *x* values, therefore, use the class midpoints as the *x* values.

Histograms can be used to show either type of distribution graphically. Frequency or relative frequency is on the vertical axis. Be sure the bars touch each other (unlike bar graphs). Increments and widths of bars should all be equal. A title should also be given to the histogram. ...

Relative frequency = frequency
                     sample size

2.31 a.

age	frequency		b. age	rel.freq.
17	1		17	0.02
18	3		18	0.06
19	16		19	0.32
20	10		20	0.20
21	12		21	0.24
22	5		22	0.10
23	1		23	0.02
24	2		24	0.04
	----------			-----------
	50			1.00

CHECKS: sum of frequencies = sample size
          sum of relative frequencies = 1.00

c.

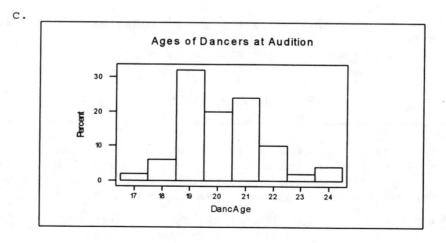

Parts of a grouped frequency distribution -

    class boundaries = the low and the high endpoints of the interval

    class width = distance from any point in one class to the same position point in the next class or the difference between the upper and lower class boundaries

    class mark = (lower boundary + upper boundary)/2, midpoint of the interval

Example: with respect to the boxed class interval

    30 - 40   form: $(40 \le x < 50)$

    40 - 50   lower class boundary = 40

    50 - 60   upper class boundary = 50

    class width = 50 - 40 = 10

    class mark = (40 + 50)/2 = 45

2.33 a. 12 - 16
     b. 2, 6, 10, 14, 18, 22, 26
     c. 4.0
     d. 0.08, 0.16, 0.16, 0.40, 0.12, 0.06, 0.02
     e.

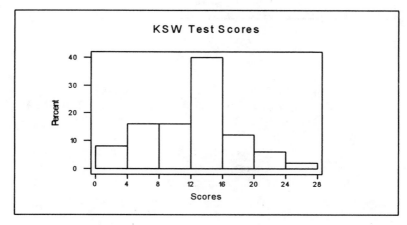

Refer to frequency distribution information before exercise 2.33 if necessary. Either class boundaries or class marks, may be used to determine increments along the horizontal axis for histograms of grouped frequency distributions.

2.35 a.

Class limits	frequency
12 - 18	1
18 - 24	14
24 - 30	22
30 - 36	8
36 - 42	5
42 - 48	3
48 - 54	2

b. class width = <u>6</u>

c. class mark = (24+30)/2

= <u>27</u>

lower class
boundary = <u>24</u>
upper class
boundary = <u>30</u>

d.

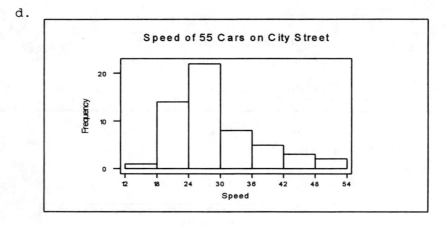

2.37 a.

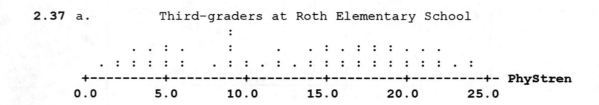

b.

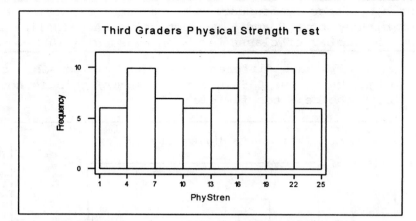

c.

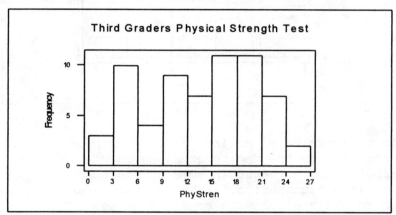

d.

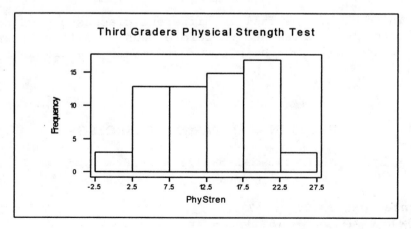

e. Each will have different results. One possible choice might be to use 0-5-10-.... However, this is not the best choice - too few classes.

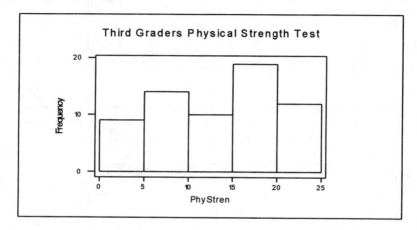

f. There seems to be an indication of a bimodal distribution. Depending on the choice of class boundaries, this bimodal appearance is weaker in some graphs than in others. A possible explanation for bimodal appearance is that the third graders are both boys and girls.

g. The data values fall in different classes depending on the class boundaries and therefore causing different grouping of the data, which in turn determines the frequency in the classes.

SECTION 2.3 MARGIN EXERCISES

2.39 $\bar{x} = \Sigma x/n = (1+2+1+3+2+1+5+3)/8 = 18/8 = \underline{2.25}$

2.40 Ranked data: 70, 72, 73, 74, 76
$d(\tilde{x}) = (n+1)/2 = (5+1)/2 = 3rd; \tilde{x} = \underline{73}$

2.41 Ranked data: 4.15, 4.25, 4.25, 4.50, 4.60, 4.60, 4.75, 4.90
$d(\tilde{x}) = (n+1)/2 = (8+1)/2 = 4.5th$; $\tilde{x}$ = <u>4.55</u>

2.42 mode = <u>2</u>

2.43 midrange = $(L+H)/2 = (0.13+10.62)/2 = 10.75/2 = 5.375$ = <u>$5.38</u>

2.44 $\overline{x} = \Sigma x/n = (9+6+7+9+10+8)/6 = 49/6 = 8.166$ = <u>8.2</u>
Ranked data: 6, 7, 8, 9, 9, 10
$d(\tilde{x}) = (n+1)/2 = (6+1)/2 = 3.5th$; $\tilde{x}$ = <u>8.5</u>
mode = <u>9</u>
midrange = $(L+H)/2 = (6+10)/2 = 16/2$ = <u>8.0</u>

SECTION 2.3 EXERCISES

NOTE: A <u>measure of central tendency</u> is a value of the variable. It is that value which locates the "average" value for a set of data. The "average" value may indicate the "middle" or the "center" or the most popular data value.

NOTATION AND FORMULAS FOR MEASURES OF CENTRAL TENDENCY

> Σx = sum of data values
> n = # of data values in the sample
> $\overline{x}$ = sample mean = $\Sigma x/n$
> $\tilde{x}$ = sample median = middle data value
> $d(\tilde{x})$ = depth or position of median = $(n + 1)/2$
> mode = the data value that occurs most often
> midrange = (highest value + lowest value)/2

NOTE: REMEMBER TO RANK THE DATA BEFORE FINDING THE MEDIAN.
  $d(\tilde{x})$ only gives the depth or position, not the value of the median. If n is even, $\tilde{x}$ is the average of the two middle values.

  See Introductory Concepts (ST*-294) for additional information about the Σ (summation) notation.

  . . .

```
OR:

MINITAB commands to find the mean and median can be found in
JES2-pp55&57 respectively.

There are no "mode" or "midrange" MINITAB commands. Using the "let"
command and  K's = constants (single numbers), the midrange can be
calculated.
                 LET K1 = (MAX(C1) + MIN(C1))/2
                 PRINT K1

*(ST denotes this manual, Statistical Tutor)
```

2.45 a. $\overline{x}$ = $\Sigma x/n$ = (2+4+7+8+9)/5 = 30/5 = <u>6.0</u>
b. d($\tilde{x}$) = (n+1)/2 = (5+1)/2 = 3rd; $\tilde{x}$ = <u>7</u>
c. <u>no mode</u>, no value repeats
d. midrange = (H+L)/2 = (9+2)/2 = 11/2 = <u>5.5</u>

2.47 {4, 5, 5, 6, 6, 6, 7, 7, 7, 7, 8, 8, 8, 9, 11}
a. $\overline{x}$ = $\Sigma x/n$ = 104/15 = 6.9333 = <u>6.9</u>
b. d($\tilde{x}$) = (n+1)/2 = (15+1)/2 = 8th; $\tilde{x}$ = <u>7</u>
c. mode = <u>7</u>
d. midrange = (H+L)/2 = (11+4)/2 = <u>7.5</u>

2.49 The midrange = (high + low)/2 = (19.1 + (-8.4))/2 = <u>5.35</u>.
That is, the average (midrange) of median metro home price
change was an increase of 5.35%.

2.51 Ranked data: 25 25 26 27 27   29 30 30 30 30
                        30 31 31 32 32   32 33 33 34 34

a. $\overline{x}$ = $\Sigma x/n$ = 601/20 = <u>30.05</u>
d($\tilde{x}$) = (n+1)/2 = (20+1)/2 = 10.5th; $\tilde{x}$ = <u>30</u>
mode = <u>30</u>
midrange = (H+L)/2 = (34+25)/2 = <u>29.5</u>

b. Police Recruits   # $ = mean, $\overline{x}$=30.05
                             •   * = midrange = 29.5
                             •   # = median = mode = 30

```
                                    •   •
        •       •           •   •   •   •       •
        •   •   •       •   •   •   •   •       •
    --+---+---+---+---+-*-+$--+---+---+---+--
     25  26  27  28  29  30  31  32  33  34
              exercise capacity (minutes)
```

c. All deal with locating the 'center' of the data. The four
values are very close because there are no mavericks and
the data is somewhat evenly distributed about the mean.

2.53 Many different answers are possible.
    a. mean; total number of cars $\Sigma x = n \cdot \overline{x}$, where n = number
       of apartments.

    b. "1.9" cannot be the median or midrange because the data
       values (number of vehicles) would all be whole numbers;
       therefore, the median or midrange would either be a whole
       number or a '0.5' number. "1.9" cannot be the mode
       because all the data values would be whole numbers.

    c. (1.9)(256) = 486.4; (0.90)(486.4) = 437.76 = <u>438 spaces</u>

2.55 a. Ranked data: 25,500   31,500   31,500   31,500   31,500
                                    35,250   36,750   37,500   39,000   54,000

       $\overline{x} = \Sigma x/n = 354,000/10 = $ <u>35,400</u>
       $d(\tilde{x}) = (n+1)/2 = (10+1)/2 = 5.5\text{rd};$ $\tilde{x} = $ <u>33,375</u>
       mode = <u>31,500</u>
       midrange = (H+L)/2 = (54,000+25,500)/2 = <u>39,750</u>
       The values of these statistics all agree with those in
       the article.

    b. The large value of 54,000 is pulling the mean and
       midrange towards the larger data value.

2.56 range = H - L = \$32.43 - \$13.15 = \$19.28; The range tells that the costs per square foot of US city office space fall within this interval from \$13.15 to \$32.43; that is a spread or difference of \$19.28.

2.57 a. The data value x = 45 is 12 units above the mean; therefore the mean must be 33.

b. The data value x = 84 is 20 units below the mean; therefore the mean must be 104.

2.58 The mean is the 'balance point' or 'center of gravity' to all the data values. Since the weights of the data values on each side of $\overline{x}$ are equal, $\Sigma(x - \overline{x})$ will give a positive amount and an equal negative amount, thereby canceling each other out.

2.59 1st: find mean, $\overline{x} = \Sigma x/n = 25/5 = 5$

$\underline{x}$	$\underline{x - \overline{x}}$	$\underline{(x - \overline{x})^2}$	
1	-4	16	$s^2 = \Sigma(x-\overline{x})^2/(n-1)$
3	-2	4	
5	0	0	$= 46/4 = \underline{11.5}$
6	1	1	
$\underline{10}$	$\underline{5}$	$\underline{25}$	
Σ 25	0	46	

2.60

$\underline{x}$	$\underline{x^2}$	
1	1	$SS(x) = \Sigma x^2 - ((\Sigma x)^2/n)$
3	9	
5	25	$= 171 - ((25)^2/5)$
6	36	$= 171 - 125 = 46$
$\underline{10}$	$\underline{100}$	
25	171	$s^2 = SS(x)/(n-1) = 46/4 = \underline{11.5}$

SECTION 2.4 EXERCISES

NOTE: A <u>measure of dispersion</u> is a value of the variable. It is that value which describes the amount of variation or spread in a data set. A small measure of dispersion indicates data that are closely grouped, whereas, a large value indicates data that are more widely spread.

MEASURES OF DISPERSION - THE SPREAD OF THE DATA

<u>Range</u> = highest value - lowest value

<u>Standard Deviation</u> - s - the average distance a data value is from the mean

$$s = \sqrt{\sum(x - \overline{x})^2 / (n - 1)}$$

Variance - s^2 - the square of the standard deviation
(i.e., before taking the square root)

For exercises 2.61-2.63, be sure that the $\sum(x - \overline{x}) = 0$.
If using a statistical calculator (one that lets you input the data points) to find the standard deviation of a sample, use the $\sigma(n-1)$ or s*x* key. $\sigma(n)$ or σx would give the population standard deviation; that is, divide by "n" instead of "n-1".

NOTE: Standard deviation and/or variance cannot be negative. This would indicate an error in sums or calculations.

See Introductory Concepts (ST-p305) for additional information about Rounding Off.

OR:

MINITAB commands to find the range and standard deviation can be found in JES2-p67.

2.61 a. range = H - L = 9 - 2 = <u>7</u>
  b. 1st: find mean, $\overline{x} = \sum x/n = 30/5 = 6$

x	x − x̄	(x − x̄)²
2	-4	16
4	-2	4
7	1	1
8	2	4
9	3	9
∑ 30	0	34

$s^2 = \sum(x-\overline{x})^2/(n-1)$

$= 34/4 = \underline{8.5}$

  c. $s = \sqrt{s^2} = \sqrt{8.5} = 2.915 = \underline{2.9}$

2.63 a. 1st: find mean, $\bar{x} = \Sigma x/n = 72/10 = 7.2$

x	$x - \bar{x}$	$(x - \bar{x})^2$
3	-4.2	17.64
5	-2.2	4.84
5	-2.2	4.84
6	-1.2	1.44
7	-0.2	0.04
7	-0.2	0.04
7	-0.2	0.04
9	1.8	3.24
10	2.8	7.84
13	5.8	33.64
Σ 72	0	73.60

$s^2 = \Sigma(x-\bar{x})^2/(n-1)$

$= 73.60/9$

$= 8.1778 = \underline{8.2}$

An _easier_ formula for s - sample standard deviation

1. Calculate "the sum of squares for x", SS(x):
$$SS(x) = \Sigma x^2 - ((\Sigma x)^2/n)$$

2. $s = \sqrt{SS(x)/(n-1)}$

This formula eliminates the problem of accumulating round-off errors.

NOTE: SS(x) is formed from the "sum of squared deviations from the mean", $\Sigma(x - \bar{x})^2$. Σx^2 is the "sum of the squared x's". $SS(x) \neq \Sigma x^2$.

---

b.

x	x^2
3	9
5	25
5	25
6	36
7	49
7	49
7	49
9	81
10	100
13	169
Σ 72	592

$SS(x) = \Sigma x^2 - ((\Sigma x)^2/n)$

$= 592 - ((72)^2/10)$

$= 592 - 518.4 = 73.6$

$s^2 = SS(x)/(n-1)$

$= 73.6/9 = 8.1778 = \underline{8.2}$

c. $s = \sqrt{s^2} = \sqrt{8.1778} = 2.8597 = \underline{2.9}$

2.65 a.

x	x^2
1554	2414916
2291	5248681
2084	4343056
4443	19740249
2884	8317456
2478	6140484
3087	9529569
3708	13749264
2510	6300100
2055	4223025
3000	9000000
2052	4210704
2550	6502500
2013	4052169
Σ 36,709	103,772,173

$SS(x) = \Sigma x^2 - ((\Sigma x)^2/n)$

$\qquad = 103,772,173 - ((36,709)^2/14)$

$\qquad = 103,772,173 - 96,253,620.07$

$\qquad = 7,518,552.929$

$s^2 = SS(x)/(n-1)$

$\qquad = 7,518,552.929/13 = 578,350.2253$

$\qquad = \underline{578,350.2}$

b. $s = \sqrt{s^2} = \sqrt{578350.2253} = 760.4934 = \underline{760.5}$

2.67 a. $n = 10$, $\Sigma x = 200,772$ $\Sigma x^2 = 5,064,438,590$
$\overline{x} = 20,077.2$ $s^2 = 114,833,221.3$ $s = 10,716.02638$

b.
```
   .    . .            .        .   .         .    :
----+---------+---------+---------+---------+---------+-Miles
   6000     12000     18000     24000     30000     36000
```

c. With a large standard deviation, there is a large spread
   to the data. This may be the case. To be sure, check the
   mean to see if it is somewhat in the center of the data.
   If it is to one side, a large standard deviation is needed
   to cover the mavericks.

2.69 a. range = H - L = 243 - 160 = $\underline{83}$

$n = 27$, $\Sigma x = 5,255$, $\Sigma x^2 = 1,031,585$

$SS(x) = \Sigma x^2 - ((\Sigma x)^2/n) = 1,031,585 - (5,255^2/27)$
$\qquad = 8806.296297$
$s^2 = SS(x)/(n-1) = 8806.296297/26 = 338.703704$

$s = \sqrt{s^2} = \sqrt{338.703704} = 18.4039 = \underline{18.4}$

-- 29 --

b. Stem-and-leaf of SchoolYr   N = 27
   Leaf Unit = 1.0

```
   range = 83
    1  ↑ 16  0
    2  |  17  5
   11  |  18  000024555
   (7) |  19  0001225        ↑    standard deviation
    9  |  20  000            |         s = 18.4
    6  |  21  166            ↓
    3  |  22  0
    2  |  23  3
    1  ↓ 24  3
```

c. Distribution is mounded, slightly skewed; range is
   approximately 4.5 standard deviations.

2.71 The statement is incorrect. The standard deviation can never
be negative. There has to be an error in the calculations or
a typographical error in the statement.

2.73 a. Command #1 -- k1 is the mean found by summing the data and
   dividing it by the number of pieces of data
   #2 -- the mean is subtracted from each data value and the
   differences are placed in column 2
   #3 -- each of the differences is squared and placed in
   column 3
   #4 -- k2 is the variance, the sum of all the squared
   differences divided by (n-1)
   #5 -- k3 is the standard deviation, found by taking the
   square root of the variance
   #6 -- results

Row	C1	C2	C3
1	7	-0.2	0.04
2	6	-1.2	1.44
3	10	2.8	7.84
4	7	-0.2	0.04
5	5	-2.2	4.84
6	9	1.8	3.24
7	3	-4.2	17.64
8	7	-0.2	0.04
9	5	-2.2	4.84
10	13	5.8	33.64

```
#7 -- results
K1        7.20000
K2        8.17778
K3        2.85968

#8 -- results
Column Standard Deviation
Standard deviation of C1      =        2.8597
```

b. let k1 = sum(c1)/count(c1) = Steps 1 & 2
   let c2 = c1 - k1 = Step 3
   let c3 = c2**2 = Step 4
   let k2 = sum(c3)/(count(c1)-1) = Step 5

c. The procedure "goes through" the data list repeatedly;
    it sums, it counts, it finds deviations, it squares
    each deviation, and so on.

SECTION 2.5 MARGIN EXERCISES

2.75 a.

x	f	xf	x^2f
0	1	0	0
1	3	3	3
2	8	16	32
3	5	15	45
4	3	12	48
Σ	20	46	128

b. $\Sigma f = 20$; $\Sigma xf = 46$;

$\Sigma x^2 f = 128$

c. x = 4; 4 is one of the possible data values
   f = 8; 8 is the number of times an 'x' value occurred
        (x = 2 in this case)
   Σf : sum of the frequencies = sample size
   Σxf : sum of the products formed by multiplying a data
        value by its frequency; the sum of the data

2.76 a. Sum of the x-column has no meaning unless each value
         occurred only once.
      b. Each data value is multiplied by how many times it
         occurred. Summing these products will give the same sum
         if all data values were listed individually.
         Note: x   f   xf
                3   5   15   xf = 15   or   3+3+3+3+3=15

2.77 $\bar{x} = \Sigma xf/\Sigma f = 46/20 = \underline{2.3}$

2.78 $SS(x) = \Sigma x^2 f - ((\Sigma xf)^2/\Sigma f)$
$\qquad = 128 - (46^2/20) = 128 - 105.8 = 22.2$

$\qquad s^2 = SS(x)/(n-1)$
$\qquad\quad = 22.2/19 = 1.16842 = \underline{1.2}$

2.79 $s = \sqrt{s^2} = \sqrt{1.16842} = 1.080935 = \underline{1.1}$

2.80

Class limits	x	f	xf	x²f
2 - 6	4	2	8	32
6 - 10	8	10	80	640
10 - 14	12	12	144	1728
14 - 18	16	9	144	2304
18 - 22	20	7	140	2800
Σ		40	516	7504

$\bar{x} = \Sigma xf/\Sigma f = 516/40 = \underline{12.9}$

$SS(x) = \Sigma x^2 f - ((\Sigma xf)^2/\Sigma f) = 7504 - (516^2/40) = 847.6$

$s^2 = SS(x)/(n-1) = 847.6/39 = 21.7333 = \underline{21.7}$

$s = \sqrt{s^2} = \sqrt{21.7333} = 4.6619 = \underline{4.7}$

SECTION 2.5 EXERCISES

> <u>MEAN AND STANDARD DEVIATION OF FREQUENCY DISTRIBUTIONS</u>
>
> <u>Mean</u> - $\bar{x} = \Sigma xf/\Sigma f$
>
> <u>Standard Deviation</u> - s
>
> $\qquad$ 1. $SS(x) = \Sigma x^2 f - ((\Sigma xf)^2/\Sigma f)$
>
> $\qquad$ 2. $s^2 = SS(x)/(n-1)$
>
> $\qquad$ 3. $s = \sqrt{s^2}$
>
> $\qquad\qquad\qquad\qquad\qquad\qquad\qquad\qquad\qquad$ . . .

NOTE: in grouped frequency distributions, the calculated statistics are approximations.

There are no* "mean of grouped data" nor "standard deviation of grouped data" MINITAB commands. Therefore, the "let" command is used to form the xf and x^2f columns and to work through the formulas to calculate the mean and standard deviation. These MINITAB commands can be found in JES2-p74.
* Grouped data techniques are for "hand" calculations. Usually a computer has all of the data and has no problem working with large sets of data.

2.81

x	f	xf	x²f
0	15	0	0
1	12	12	12
2	26	52	104
3	14	42	126
4	4	16	64
6	2	12	72
Σ	73	134	378

$\bar{x} = \Sigma xf / \Sigma f = 134/73 = 1.836 = \underline{1.8}$

$SS(x) = \Sigma x^2 f - ((\Sigma xf)^2 / \Sigma f)$
$= 378 - (134^2/73)$
$= 378 - 245.97260 = 132.0274$

$s^2 = SS(x)/(n-1)$
$= 132.0274/72 = 1.8337 = \underline{1.8}$

$s = \sqrt{s^2} = \sqrt{1.8337} = 1.354 = \underline{1.4}$

In exercises 2.83 through 2.87, the calculated means, variances, and standard deviations of the grouped frequency distributions will be approximations. This is due to the use of the class marks versus the actual data values.

For example, suppose the class limits for a particular class are 0-6, and that 5 data values fall in that class interval. The class mark of 3 would be used in the calculations, thereby treating all 5 data values as 3s, when they each could be any numbers from 0 through 6 (even all 0's).

2.83

Class limits	x	f	xf	x²f
3 - 6	4.5	2	9	40.50
6 - 9	7.5	10	75	562.50
9 - 12	10.5	12	126	1323.00
12 - 15	13.5	9	121.5	1640.25
15 - 18	16.5	7	115.5	1905.75
	Σ	40	447.0	5472.00

-- 33 --

$$\overline{x} = \sum xf/\sum f = 447.0/40 = 11.175 = \underline{11.2}$$

$$SS(x) = \sum x^2 f - ((\sum xf)^2/\sum f) = 5472 - (447^2/40) = 476.775$$
$$s^2 = 476.775/39 = 12.225 = \underline{12.2}$$

$$s = \sqrt{s^2} = \sqrt{12.225} = 3.496 = \underline{3.5}$$

2.85

Class limits	x	f	xf	x^2f
00.00-10.00	5.00	2	10.00	50.00
10.00-20.00	15.00	8	120.00	1800.00
20.00-30.00	25.00	7	175.00	4375.00
30.00-40.00	35.00	2	70.00	2450.00
40.00-50.00	45.00	1	45.00	2025.00
	$\sum$	20	420.00	10,700.00

$$\overline{x} = \sum xf/\sum f = 420.00/20 = \underline{21.00}$$

$$SS(x) = \sum x^2 f - ((\sum xf)^2/\sum f) = 10,700.00 - (420.00^2/20)$$
$$= 1880.00$$
$$s^2 = SS(x)/(n-1) = 1880.00/19 = 98.9474$$

$$s = \sqrt{s^2} = \sqrt{98.9474} = 9.9472 = \underline{9.95}$$

2.87

Class limits	x	f	xf	x^2f
12 - 18	15	1	15	225
18 - 24	21	14	294	6174
24 - 30	27	22	594	16038
30 - 36	33	8	264	8712
36 - 42	39	5	195	7605
42 - 48	45	3	135	6075
48 - 54	51	2	102	5202
	$\sum$	55	1599	50,031

$$\overline{x} = \sum xf/\sum f$$
$$= 1599/55$$
$$= 29.0727 = \underline{29.1}$$

$$SS(x) = \sum x^2 f - ((\sum xf)^2/\sum f) = 50,031 - (1599^2/55)$$
$$= 3543.709091$$
$$s^2 = SS(x)/(n-1) = 3543.709091/54 = 65.6242$$

$$s = \sqrt{s^2} = \sqrt{65.6242} = 8.100879 = \underline{8.1}$$

OTHER MEASURES OF CENTRAL TENDENCY FOR FREQUENCY DISTRIBUTIONS

NOTE: Data are already ranked.

Median - $\tilde{x}$ - find the depth and count down the frequency column until you include that position number. This is the median class. In an ungrouped frequency distribution, the median equals the x value of that class. In a grouped distribution, the data must be ranked in that particular class, then count to the appropriate position. If the original data are not given, use the class mark.

Mode - class mark of the highest frequency class

Modal Class - interval bounded by the class boundaries of the class with the highest frequency

Midrange = (highest value + lowest value)/2. If the original data is not given, use the lowest class boundary and the highest class boundary from the entire distribution, or the lowest and highest class marks.

2.89 a. n = Σf = 89, Σxf = 54,100;
$\overline{x}$ = Σxf/Σf = 54,100/89 = 607.865 = $608

b. In order for $734 to be the "mean" of the sample there would need to be many large values of x [much larger than $1500]. It would appear that 734 is not the mean.

c. median = $450. [The median is in the group between 301 and 600.]

d. The $734 can not be the median, not even if all of the "Didn't Know" spent over $1000 [the median is still in the group 301 to 600].

e. Mode of the frequency distribution is $450.

f. The mode is most likely a value in one of the two largest classes, thus between $1 and $600. However, since the mode is the single value that occurs most often, it could be that $734 occurred maybe four times and no other value repeated more than three times. It could be the mode.

g. Yes, the average of $734 could be the midrange. If the largest amount reported was $1468, then the midrange would be exactly $734. It seems more likely that one person would report $1468 as an amount spent than several would report exactly $734.

2.91 91 is in the 44th position from the Low value of 39
     91 is in the 7th position from the High value of 98

2.92 $nk/100 = (50)(20)/100 = 10.0$; therefore $d(P_{20}) = 10.5$th from L
      $P_{20} = (64+64)/2 = \underline{64}$
     $nk/100 = (50)(35)/100 = 17.5$; therefore $d(P_{35}) = 18$th from L
      $P_{35} = \underline{70}$

2.93 $nk/100 = (50)(20)/100 = 10.0$; therefore $d(P_{80}) = 10.5$th from H
      $P_{80} = (88+89)/2 = \underline{88.5}$
     $nk/100 = (50)(5)/100 = 2.5$; therefore $d(P_{95}) = 3$rd from H
      $P_{95} = \underline{95}$

2.94 The distribution needs to be symmetric about the mean.

2.95

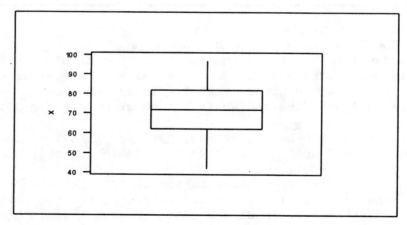

2.96 $z = (x - \text{mean})/\text{st.dev.}$

     for $x = 92$, $z = (92 - 72)/12 = \underline{1.67}$
     for $x = 63$, $z = (63 - 72)/12 = \underline{-0.75}$

NOTE: A <u>measure of position</u> is a value of the variable. It is that value which divides the set of data into two groups: those data smaller in value than the measure of position, and those larger in value than the measure of position.

To find any measure of position:

    1. Rank the data - <u>DATA MUST BE RANKED LOW TO HIGH</u>

    2. Determine the depth or position in two separate steps:
       a. Calculate $nk/100$, where n = sample size,
                            k = desired percentile
       b. Determine $d(P_k)$:
            If $nk/100$ = integer $\Rightarrow$ add .5 (value will be half-
                                   way between 2 integers)
            If $nk/100$ = decimal $\Rightarrow$ round up to the nearest
                                 whole number
    3. Locate the value of P_k

REMEMBER:
$Q_1 = P_{25}$ = 1st quartile - 25% of the data lies below this value
$Q_2 = P_{50} = \tilde{x}$ = 2nd quartile - 50% of the data lies below this value
$Q_3 = P_{75}$ = 3rd quartile - 75% of the data lies below this value

2.97 Ranked data:
    2.6 2.7 3.4 3.6 3.7   3.9 4.0 4.4 4.8 4.8
    4.8 5.0 5.1 5.6 5.6   5.6 5.8 6.8 7.0 7.0

    a. $nk/100$ = (20)(25)/100 = 5.0; therefore $d(P_{25})$ = 5.5th
        $Q_1 = P_{25}$ = (3.7 + 3.9)/2 = <u>3.8</u>
        $nk/100$ = (20)(75)/100 = 15.0; therefore $d(P_{75})$ = 15.5th
        $Q_3 = P_{75}$ = (5.6 + 5.6)/2 = <u>5.6</u>

    b. midquartile = $(Q_1 + Q_3)/2$ = (3.8 + 5.6)/2 = <u>4.7</u>

    c. $nk/100$ = (20)(15)/100 = 3.0; therefore $d(P_{15})$ = 3.5th
        P_{15} = (3.4+3.6)/2 = <u>3.5</u>

        $nk/100$ = (20)(33)/100 = 6.6; therefore $d(P_{33})$ = 7th
        P_{33} = <u>4.0</u>

        $nk/100$ = (20)(90)/100 = 18.0; therefore $d(P_{90})$ = 18.5th
        P_{90} = (6.8+7.0)/2 = <u>6.9</u>

2.99   a. $d(\tilde{x}) = (n+1)/2 = (35+1)/2 = 18$th;   median, $\tilde{x}$ = __33.0__

b. midrange = $(H+L)/2 = (39.5+30.1)/2 = $ __34.8__

c. $nk/100 = (35)(25)/100 = 8.75$; therefore $d(P_{25})$ = 9th
   $Q_1 = P_{25} = 31.3$

   $nk/100 = (35)(75)/100 = 26.25$; therefore $d(P_{75})$ = 27th
   $Q_3 = P_{75} = 36.0$

   midquartile = $(Q_1 + Q_3)/2 = (31.3 + 36.0)/2 = $ __33.65__

d. 30.1   31.3   33.0   36.0   39.5
   _____

   L Q_1 $\tilde{x}$ Q_3 H

e.

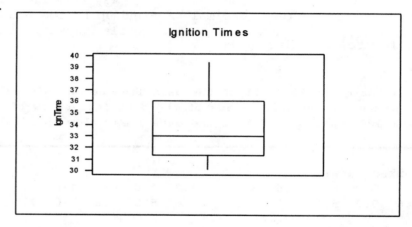

z is a measure of position. It gives the number of standard
deviations a piece of data is from the mean. It will be <u>positive</u>
if x is to the <u>right of the mean</u> (larger than the mean) and
<u>negative</u>, if x is to the <u>left of the mean</u> (smaller than the mean).
Keep 2 decimal places. (hundredths)

$\qquad$ z = (x - mean)/st. dev. $\qquad$ z = $(x - \overline{x})/s$

2.101 z = (x - mean)/st.dev.

a. for x = 54,   z = $(54 - 50)/4.0 = $ __1.0__
b. for x = 50,   z = $(50 - 50)/4.0 = $ __0.0__
c. for x = 59,   z = $(59 - 50)/4.0 = $ __2.25__
d. for x = 45,   z = $(45 - 50)/4.0 = $ __-1.25__

2.103 If z = (x - mean)/st.dev; then   x = (z)(st.dev) + mean

for z = 1.8,   x = (1.8)(100) + 500 = <u>680</u>

2.105 a. 152 is one and one-half standard deviations above the mean
b. the score is 2.1 standard deviations below the mean
c. the number of standard deviations from the mean

2.107 for A: z = (85 - 72)/8 = 1.625
for B: z = (93 - 87)/5 = 1.2
Therefore, <u>A has the higher relative position</u>.

SECTION 2.7 MARGIN EXERCISES

2.109 From 175 through 225 words, inclusive.

2.110 $1 - (1/k^2) = 1 - (1/4^2) = 1 - (1/16) = 15/16 = 0.9375$;
<u>at least 93.75%</u>

2.111 a. 50%
b. ≈68%
c. ≈84%

SECTION 2.7 EXERCISES

<u>Chebyshev's theorem</u> applies to any shape distribution.
    At least 75% of the data lies within 2 standard deviations
    of the mean.
    At least 89% of the data lies within 3 standard deviations
    of the mean.

<u>The empirical rule</u> applies to a normal distribution.
    Approximately 68% of the data lies within 1 standard
    deviation of the mean.
    Approximately 95% of the data lies within 2 standard
    deviations of the mean.
    Approximately 99.7% of the data lies within 3 standard
    deviations of the mean.

2.113 a. 97.6 is 2 standard deviations above the mean $\{z = (97.6-84.0)/6.8 = 2.0\}$, therefore <u>2.5%</u> of the time more than 97.6 hours will be required.

b. 95% of the time the time to complete will fall within 2 standard deviations of the mean, that is $84.0 \pm 2(6.8)$ or from <u>70.4</u> to <u>97.6</u> hours.

Chebyshev's theorem

At least $\left(1 - \dfrac{1}{k^2}\right)$% of the data lies within k standard deviations of the mean. (k > 1)

2.115 a. at least 75%   b. approximately 95%

2.117 a.

x	f	xf	x²f
1	7	7	7
2	10	20	40
3	22	66	198
4	8	32	128
5	7	35	175
6	2	12	72
7	3	21	147
8	0	0	0
9	1	9	81
Σ	60	202	848

b. $\bar{x} = \Sigma xf/\Sigma f$
   $= 202/60 = 3.367 = \underline{3.4}$

$SS(x) = \Sigma x^2 f - ((\Sigma xf)^2/\Sigma f)$
       $= 848 - (202^2/60)$
       $= 167.9333$

$s = \sqrt{s^2} = \sqrt{167.9333/59}$
  $= \sqrt{2.8463} = 1.687 = \underline{1.7}$

c. $\bar{x} - s = 3.4 - 1.7 = \underline{1.7}$; $\bar{x} + s = 3.4 + 1.7 = \underline{5.1}$

d. <u>47</u>, <u>78%</u> (47/60)

e. $\bar{x} - 2s = 3.4 - 2(1.7) = \underline{0.0}$; $\bar{x} + 2s = 3.4 + 2(1.7) = \underline{6.8}$

f. <u>56</u>, <u>93%</u> (56/60)

g. $\bar{x} - 3s = 3.4 - 3(1.7) = \underline{-1.7}$; $\bar{x} + 3s = 3.4 + 3(1.7) = \underline{8.5}$

h. <u>59</u>, <u>98.3%</u> (59/60)

i. 93% is at least 75% and 98.3% is at least 89%; both agree with Chebyshev's Theorem.

j. 78%, 93% and 98.3% are not approximately equal to the 68%, 95%, and 99.7% of the empirical rule. It appears that more data than expected occurs within one standard deviation of $\bar{x}$; the distribution is probably not normal.

Helpful hint for use when expecting to count data on histogram:

Menu command: While on the Histogram dialogue box,
     Select: **Annotations > Data Labels...**
     Select: **Show data labels**
This will direct the computer to print the frequency of each class above its corresponding bar.

2.119 a.

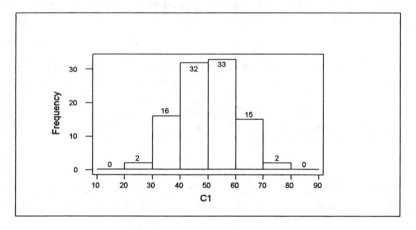

Within one standard deviation, 40 to 60, is 33 + 32 or 65 of the 100 data. 65%
Within two standard deviations, 30 to 70, is 16+32+33+15 or 96 of the 100 data. 96%
Within three standard deviations, 20 to 80, is all 100 of the data, or 100%.

The above results are extremely close to what the empirical rule claims will happen.

b, c, d. Not all sets of 100 data will result in percentages this close. However, expect very similar results to occur most of the time.

SECTION 2.8 MARGIN EXERCISES

2.121 Yes, if all 8 employees earned $300 each, the mean would be $405.56 and if all 8 employees earned $350 each, the mean would be $450. $430 falls within this interval.

Or, if the mean of 9 employees is 430, then the total is 3870. The 8 employees then would need to make 2620 (3870-1250) and their average earnings would be 2620/8 or 327.50, which is within the interval.

2.122 Half of anything is 50%. Strange, but according to this graph, 32% is more than half.

The circle graph could be corrected by including all 100% of the responses. Another correction, still misleading but not as obvious would be to use a bar graph which shows the proportional size of the categories, but does not necessarily show the whole situation.

2.123 Figure 1, by cutting off the bottom of the graph, emphasizes the variation between the numbers with no regards to the relative size of the numbers. Figure 2 shows the relative size of the numbers being presented and basically shows that the difference between the values is relatively small compared to their size.

2.124 The graph would be a bar graph with each bar showing the relative frequency of fatal injuries for each separate age class.

SECTION 2.8 EXERCISES

2.125 The class width is not uniform.

2.127 a. Mean increased; when one data increases, the sum increases.

b. Median is unchanged; the median is affected only by the middle value(s).

c. Mode is unchanged.

d. Midrange increased; an increase in either extreme value increases the sum H+L.

e. Range increased; difference between high and low values increased.

f. Variance increased; data are now more spread out.

g. Standard deviation increased; data are now more spread out.

2.129 Data summary: $n = 8$, $\sum x = 36.5$, $\sum x^2 = 179.11$

a. $\bar{x} = \sum x/n = 36.5/8 = 4.5625 = \underline{4.56}$

b. $s = \sqrt{[\sum x^2 - ((\sum x)^2/n)]/(n-1)}$
   $= \sqrt{[179.11 - (36.5^2/8)]/7}$
   $= \sqrt{1.79696} = 1.3405 = \underline{1.34}$

c. These percentages seem to average very closely to 4%.

2.131 Data summary: $n = 118$, $\sum x = 2364$

a. $\bar{x} = \sum x/n = 2364/118 = 20.034 = \underline{20.0}$

b. $d(\tilde{x}) = (n+1)/2 = (118+1)/2 = 59.5$th;
   $\tilde{x} = (17+17)/2 = \underline{17}$

c. mode = $\underline{16}$

d. $nk/100 = (118)(25)/100 = 29.5$; therefore $d(P_{25}) = 30$th
   $Q_1 = P_{25} = \underline{15}$

   $nk/100 = (118)(75)/100 = 88.5$; therefore $d(P_{75}) = 89$th
   $Q_3 = P_{75} = \underline{21}$

e. $nk/100 = (118)(10)/100 = 11.8$; therefore $d(P_{10}) = $ 12th
$P_{10} = \underline{14}$

$nk/100 = (118)(95)/100 = 112.1$; therefore $d(P_{95}) = 113$
$P_{95} = \underline{43}$

2.133 Data: 1.4 1.7 1.1 2.4 2.5   3.5 3.0 3.4 3.1 4.4
       5.5 5.7 5.8 6.8 6.8   6.2 6.8 6.6 7.5 9.4

Data summary: $n = 20$, $\Sigma x = 93.6$, $\Sigma x^2 = 541.56$

$\overline{x} = \Sigma x/n = 93.6/20 = \underline{4.68}$

$SS(x) = \Sigma x^2 - ((\Sigma x)^2/n)$
$= 541.56 - (93.6^2/20) = 103.512$

$s^2 = SS(x)/(n-1) = 103.512/19 = 5.448$

$s = \sqrt{s^2} = \sqrt{5.448} = 2.334 = \underline{2.33}$

$\overline{x} \pm 2s = 4.68 \pm 2(2.33) = 4.68 \pm 4.66$ or 0.02 to 9.34
$\underline{95\%}$ (19 of the 20) of the data is within two standard
deviations of the mean.

2.135

x	f	xf
3	75	225
6	150	900
8	30	240
9	50	450
12	70	840
14	300	4200
15	400	6000
16	1050	16800
17	750	12750
18	515	9270
19	120	2280
20	60	1200
Σ	3570	55,155

a. $\overline{x} = \Sigma xf/\Sigma f = 55,155/3570$
    $= 15.449 = \underline{15.4}$

b. $d(\tilde{x}) = (\Sigma f+1)/2$
    $= (3570+1)/2 = 1785.5$th
   median $= \underline{16}$

c. mode $= \underline{16}$

d. midrange $= (H+L)/2$
         $= (3+20)/2 = \underline{11.5}$

2.137 a.   Number of Persistant Disagreements

```
                    :        .        :
                    :  S     .        :      .
               ------:-------:--------:-----:--S--:----------:----
               :     :       :        :     :    :          :
          :    :     :        :       :     :    :     :          :
     :    :    :     :        :       :     :    :     :     .    :
    +---------+---------+---------+---------+---------+---------+-------
   0.0       2.0       4.0      ˣ 6.0       8.0      10.0
                                no.items
```

b. $d(\tilde{x}) = (n+1)/2 = (66+1)/2 = 33.5\text{th};$ median = <u>5</u>

x	f	xf	x²f
0	2	0	0
1	2	2	2
2	4	8	16
3	10	30	90
4	7	28	112
5	9	45	225
6	8	48	288
7	11	77	539
8	7	56	448
9	3	27	243
10	1	10	100
11	2	22	242
Σ	66	353	2305

c. $\overline{x} = \Sigma xf/\Sigma f = 353/66$
$\qquad = 5.34848 = \underline{5.3}$

d. $SS(x) = \Sigma x^2 f - ((\Sigma xf)^2/\Sigma f)$
$\qquad = 2305 - (353^2/66)$
$\qquad = 416.98485$
$s = \sqrt{SS(x)/(n-1)}$
$\quad = \sqrt{416.98485/65}$
$\quad = \sqrt{6.41515} = 2.5328 = \underline{2.5}$

e. & f. On graph in (a).

2.139 a.

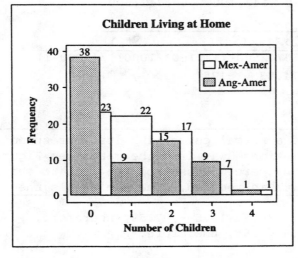

b.

x	f	xf	x²f
0	23	0	0
1	22	22	22
2	17	34	68
3	7	21	63
4	1	4	16
Σ	70	81	169

$\overline{x} = \Sigma xf/\Sigma f = 81/70$
$= 1.157 = \underline{1.2}$

$s = \sqrt{[\Sigma x^2 f - ((\Sigma xf)^2/\Sigma f)]/(\Sigma f - 1)}$
$= \sqrt{[169 - (81^2/70)]/(70-1)}$
$= \sqrt{1.09089} = 1.0445 = \underline{1.0}$

c.

x	f	xf	x²f
0	38	0	0
1	9	9	9
2	15	30	60
3	9	27	81
4	1	4	16
Σ	72	70	166

$\overline{x} = \Sigma xf/\Sigma f = 70/72$
$= 0.972 = \underline{1.0}$

$s = \sqrt{[\Sigma x^2 f - ((\Sigma xf)^2/\Sigma f)]/(\Sigma f - 1)}$
$= \sqrt{[166 - (70^2/72)]/(72-1)}$
$= \sqrt{1.3795} = 1.1745 = \underline{1.2}$

d. Yes the two distributions seem to be different. The histograms are quite different for the first three classes. Also, there is a 20% difference in the values of both the means and the standard deviations.

2.141

x	f	xf	x²f
0	87	0	0
50	544	27,200	1,360,000
150	165	24,750	3,712,500
350	157	54,950	19,232,500
750	47	35,250	26,437,500
Σ	1000	142,150	50,742,500

$\overline{x} = \Sigma xf/\Sigma f$
$= 142,150/1000$
$= 142.15 = \underline{\$142}$

$s = \sqrt{[\Sigma x^2 f - ((\Sigma xf)^2/\Sigma f)]/(\Sigma f - 1)}$
$= \sqrt{[50742500 - (142150^2/1000)]/(1000-1)}$
$= 174.83 = \underline{\$175}$

Draw a diagram of a normal curve with its corresponding percentages for standard deviations away from the mean (see Figure 2-33 in JES2-p90). Add the percentages from the left to the right, until the desired z-value is reached. The sum equals the percentile.

2.143 a. $2.5 + 13.5 + 34 + 34 + 13.5 = 97.5\%$; therefore, $\underline{P_{98}}$
   b. $2.5 + 13.5 = 16\%$; therefore, $\underline{P_{16}}$

2.145 x = (z)(st.dev) + mean

Sit-ups: x = (-1)(12) + 70 = <u>58</u>
Pull-ups: x = (-1.3)(6) + 8 = 0.2 = <u>0</u>
Shuttle Run: x = (0)(0.6) + 9.8 = <u>9.8</u>
50 yd. dash: x = (1)(.3) + 6.6 = <u>6.9</u>
Softball: x = (0.5)(16) + 173 = <u>181</u>

2.147 a. at least 75%   b. at least 89%

2.149 The empirical rule states that 99.7% of a normal
distribution is between the z-scores of -3 and +3.

2.151 Data summary: n = 8, Σx = 31,825, Σx^2 = 126,894,839

a. $\overline{x}$ = $\Sigma x/n$ = 31,825/8 = <u>3978.1</u>

b. SS(x) = Σx^2 - ((Σx)^2/n)
        = 126,894,839 - (31,825^2/8) = 291,010.88

s^2 = SS(x)/(n-1) = 291,010.88/7 = 41,572.982

s = $\sqrt{s^2}$ = $\sqrt{41572.98}$ = <u>203.9</u>

c. $\overline{x}$ ± 2s = 3978.1 ± 2(203.9)
          = 3978.1 ± 407.8   or   <u>3570.3</u> <u>to</u> <u>4385.9</u>

2.153 a.

```
          . .
  .... ..:::       .        .   .                    .
--+---------+---------+---------+---------+---------+---MedPr95
    .              .
  : . :  .. :        . : .
--+---------+---------+---------+---------+---------+---MedPr94
 70000    105000    140000    175000    210000    245000
```

b. 1995 median prices overall were about the same as 1994,
   except for the three high values in 1995.

2.155 a.

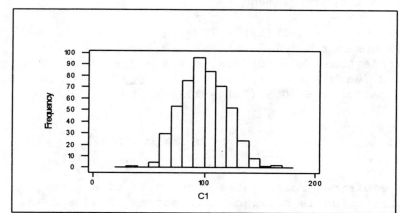

b.

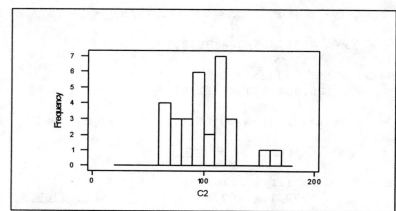

c.

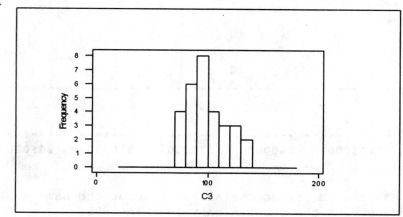

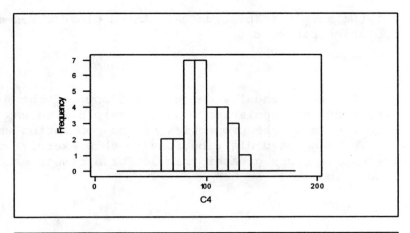

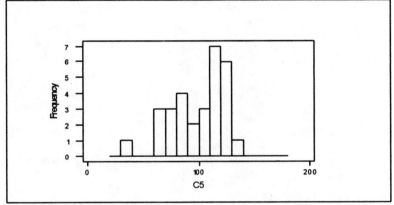

d.

Variable	N	Mean	Median	StDev
popul.	500	99.817	98.778	19.973
sample1	30	100.35	96.04	24.14
sample2	30	99.80	95.23	18.13
sample3	30	97.53	93.74	17.41
sample4	30	99.89	103.89	23.65

Variable	Min	Max	Q1	Q3
popul.	39.676	167.592	85.228	113.094
sample1	65.74	162.63	84.14	113.62
sample2	74.57	139.59	85.26	112.11
sample3	65.21	133.04	87.73	111.03
sample4	39.68	130.95	81.53	118.66

NOTE: With MINITAB, the DESCribe command gives additional descriptive statistics.

e. Yes, the sample statistics calculated closely resemble the population parameters

2.157 Samples of size 30 usually demonstrated some of the properties of the population. As the sample size was increased, more of the properties of the population were shown. The suggested distributions in this exercise seem to require sample sizes greater than 30 for a closer match to the population.

CHAPTER 3 ∇   DESCRIPTIVE ANALYSIS AND PRESENTATION OF BIVARIATE DATA

Chapter Preview

Chapter 3 deals with the presentation and analysis of bivariate (two variables) data. There are three main categories of bivariate data.

1.   Two Qualitative Variables

This type of data is best presented in a contingency table and/or bar graph. Variations of the contingency table are given in Section 1 of Chapter 3.

2.   One Qualitative Variable and One Quantitative Variable

This type of data can be presented and/or summarized in table form or graphically. More statistical techniques are available because of the one quantitative variable. Dot plots, box plots, and stem-and-leaf diagrams can represent the data for each different value of the qualitative variable.

3.   Two Quantitative Variables

Initially, this type of data is best presented in a scatter diagram. If a relationship seems to exist, based on the scatter plot, then linear correlation and regression techniques will be performed.

SECTION 3.1 MARGIN EXERCISES

3.1

	Very Good	Good	Other	Marginal total
Husband	23%	17.3%	9.7%	50%
Wife	24%	17.7%	8.3%	50%
Marginal total	47.0%	35.0%	18.0%	100%

3.2

	Very Good	Good	Other	Marginal total
Husband	46%	34.6%	19.4%	100%
Wife	48%	35.4%	16.6%	100%
Marginal total	47.0%	35.0%	18.0%	100%

(3.2 continued)
The table shows the distribution of ratings for husband and wife separately. For example, 46% of husbands rate their spouse as a "very good" love partner while 48% of wives rate their spouse as a "very good" love partner.

3.3

	Very Good	Good	Other	Marginal total
Husband	48.9%	49.4%	53.9%	50%
Wife	51.1%	50.6%	46.1%	50%
Marginal total	100%	100%	100%	100%

The table shows the distribution of husbands and wives for each of the ratings. For example, for the rating "good", 49.4% of the responses were from husbands and 50.6% from wives.

3.4

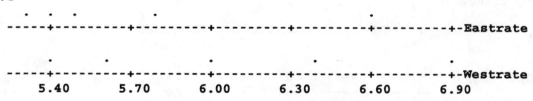

East coast cities: $\overline{x} = \Sigma x/n = 28.6/5 = 5.72$
$\qquad\qquad\qquad d(\tilde{x}) = (n+1)/2 = (5+1)/2 = 3rd;\ \tilde{x} = 5.5$
West coast cities: $\overline{x} = \Sigma x/n = 30.3/5 = 6.06$
$\qquad\qquad\qquad d(\tilde{x}) = (n+1)/2 = (5+1)/2 = 3rd;\ \tilde{x} = 6.0$

3.5 The input variable most likely would be height. Based on height, weight is often predicted or given in a range of acceptable values depending on the size of a person's frame.

3.6

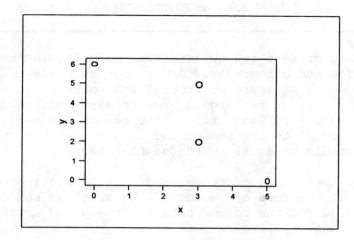

3.7 a.

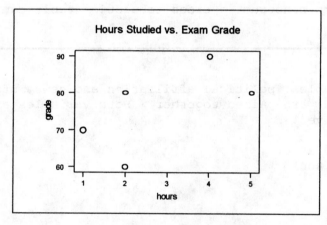

b. As hours studied increased, there seems to be a trend for the exam grades to also increase.

Exercises 3.8-3.10 present two qualitative variables in the form of contingency tables and bar graphs. A contingency table is made up of rows and columns. Rows are horizontal and columns are vertical. Adding across the rows gives marginal row totals. Adding down the columns gives marginal column totals. The sum of the marginal row totals should be equal to the sum of the marginal column totals, which in turn, should be equal to the sample size.

Before answering any questions concerning data in contingency tables, add all of the rows and columns. Be sure the sum of the row totals = the sum of the column totals = the grand total. Now you are ready to answer all questions easily.

OR:

MINITAB commands to construct a cross-tabulation table can be found in JES2-p115.

3.9 a. 3000
     b. Two variables, political affiliation and news information preferred, are paired together. Both variables are qualitative.
     c. 950
     d. 50% [1500/3000]
     e. 25% [200/800]

Exercise 3.11 demonstrates the statistical methods that can be used on "one qualitative, one quantitative" type data. Be sure to split the data based on the qualitative variable. The effect is a side-by-side comparison of the quantitative variable for each different value of the qualitative variable.

...

```
OR:

MINITAB commands to construct multiple dotplots or boxplots can be
found in JES2-p117.  Likewise multiple stem-and-leaf displays can
be constructed one for each value of the qualitative variable.
This will allow for a "side-by-side" comparison of the data.
Numerical codes (numbers) are assigned to the qualities.

   STEM C1;                    Separate stem-and-leaf displays (one
   INCRement = 1;              for each qualitative variable in
   BY C2.                      column 2) are constructed for the
                               quantitative data in column 1.  The
                               increment command specifies the
                               increment between succeeding stems.
```

3.11 a.

```
MinDepos
5    .                    .              . . : :.. :
  +---------+---------+---------+---------+---------+-------Yield
MinDepos                                  .       .
10                              . ::... :: :   .
  +---------+---------+---------+---------+---------+-------Yield
MinDepos                                 .
25                         .      . :..:.  .   .   . .
  +---------+---------+---------+---------+---------+-------Yield
  3.90      4.20      4.50      4.80      5.10      5.40
```

    b. Rates for minimum deposits of:

	$500	$1000	$2500
High	5.26	5.45	5.50
Q_3	5.19	5.32	5.24
$\tilde{x}$	5.09	5.15	5.10
Q_1	4.955	5.09	5.00
Low	4.00	5.00	4.75

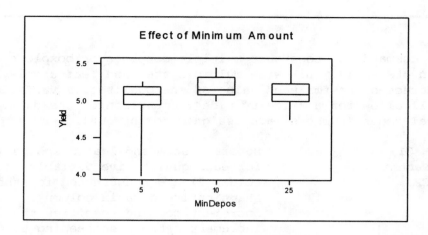

c. The middle 50% of the three sets are fairly similar. The
   $500 set has two extremely low values making for a wider
   skewed distribution. Overall, minimum deposit appears to
   have little effect on interest rate.

3.13 a.   Heights of World Cup Soccer Players

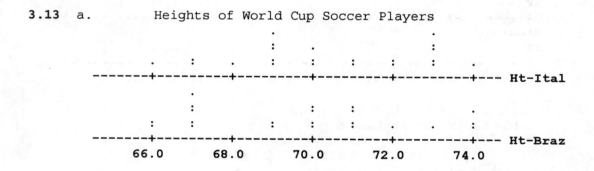

Weights of World Cup Soccer Players

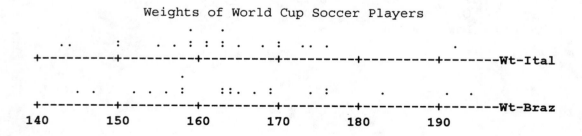

Ages of World Cup Soccer Players

```
                                     .
                          .     :  .
           :       .  .: : :     :  .    . :
     -----+---------+---------+---------+---------+---  Age-Ital
                          .           : : .
                  .       : : . .: : : : .
           .              : : . .: : : : .      .
     -----+---------+---------+---------+---------+---  Age-Braz
         15.0      20.0      25.0      30.0      35.0
```

b. No, both height and weight had the same basic spread of
   data with peaks in the same locations. The only
   difference was the one "young" Brazil player, otherwise
   the ages were also distributed pretty evenly per team.

c. The data can not be paired between the two teams.

Exercises 3.12c and 3.14-3.20 demonstrate the numerical approach
that can be taken now that we have two quantitative variables. A
scatter diagram is the first tool we use in determining whether a
linear relationship exists between the two variables. Decide which
variable is to be predicted. This variable will be the dependent
variable.* Let x be equal to the independent variable (input
variable) and y be equal to the dependent variable (output
variable).

How to construct a scatter diagram:

   1. Find the range of the x values and the range of the y values.

   2. Based on these, choose your increments for the x-axis
      (horizontal axis) and then for the y-axis (vertical axis).
      They will not always be the same.

   3. Each point on the scatter diagram is made up of an ordered pair
      (x,y). (x,y) is plotted at the point that is x units on the
      x-axis and y units on the y-axis.

   4. Label both axes and give a title to the diagram.

*(ex. the age of a car and the price of a car; price would be the
dependent variable if we wish to predict the price of a car based
on its age)   ...

OR:

MINITAB commands to construct a scatter diagram can be found in
JES2-p118. Possible additional subcommands: GRID 1;
                                                 GRID 2; will draw
in horizontal and vertical lines grid lines at labeled intervals
along the axes. These are sometimes helpful in 'reading' the
graph.

NOTE: When the data contains two or more identical ordered pairs,
they appear as only one ordered pair on the MINITAB scatter
diagram.

3.15

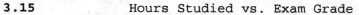

Hours Studied vs. Exam Grade

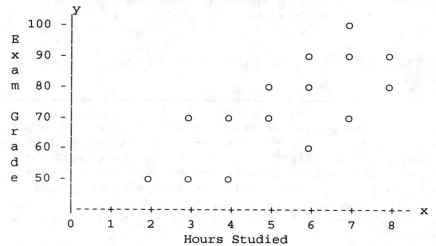

3.17

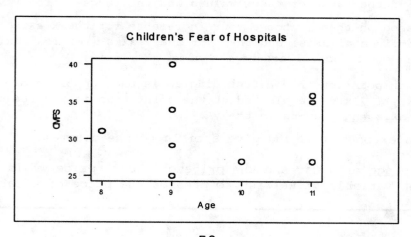

3.19 a.

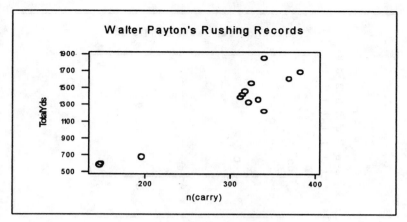

b. As the number of carries increased so did the number of total yards. There are also two separate groups of data points.

c. Two of the "low" group of three ordered pairs come from the first year and last year of his career. The other ordered pair of these three could be a year when a serious injury restricted his playing time.

SECTION 3.2 MARGIN EXERCISES

3.21 a.

x	y	x²	xy	y²
2	80	4	160	6400
5	80	25	400	6400
1	70	1	70	4900
4	90	16	360	8100
2	60	4	120	3600
14	380	50	1110	29,400

$SS(x) = \sum x^2 - ((\sum x)^2/n) = 50 - (14^2/5) = \underline{10.8}$

$SS(y) = \sum y^2 - ((\sum y)^2/n) = 29,400 - (380^2/5) = \underline{520}$

$SS(xy) = \sum xy - ((\sum x \cdot \sum y)/n) = 1110 - (14 \cdot 380/5) = \underline{46}$

b. $r = SS(xy)/\sqrt{SS(x) \cdot SS(y)} = 46/\sqrt{10.8 \cdot 520} = 0.6138 = \underline{0.61}$

3.22 a.

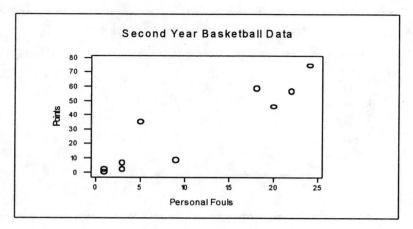

b.

x	y	x²	xy	y²
1	2	1	2	4
24	75	576	1800	5625
1	0	1	0	0
18	59	324	1062	3481
9	9	81	81	81
3	7	9	21	49
5	35	25	175	1225
20	46	400	920	2116
1	0	1	0	0
3	2	9	6	4
22	57	484	1254	3249
107	292	1911	5321	15,834

$SS(x) = \sum x^2 - ((\sum x)^2/n) = 1911 - (107^2/11) = \underline{870.1818}$

$SS(y) = \sum y^2 - ((\sum y)^2/n) = 15,834 - (292^2/11) = \underline{8082.72727}$

$SS(xy) = \sum xy - ((\sum x \cdot \sum y)/n) = 5321 - (107 \cdot 292/11) = \underline{2480.63636}$

$$r = SS(xy)/\sqrt{SS(x) \cdot SS(y)} = 2480.63636/\sqrt{870.1818 \cdot 8082.72727}$$
$$= \underline{0.94}$$

c. No, a player that scores many points is most likely in the game more, therefore more likely to have more fouls. A strong correlation coefficient indicates a strong linear (straight line) relationship not necessarily a cause and effect situation.

3.23   Estimate r to be near -0.75
Estimate r to be near 0.00
Estimate r to be near +0.75

Based on the scatter diagram, we might suspect whether or not a linear relationship exists. If the y's increase linearly as the x's increase, there exists a relationship called <u>positive correlation</u>. If the y's decrease linearly as the x's increase, there exists a relationship called <u>negative correlation</u>. The measure of the strength of this linear relationship is denoted by r, the coefficient of linear correlation.

<u>r = linear correlation coefficient</u>

1. r has a value between -1 and +1, i.e. $-1 \leq r \leq +1$

2. r = -1 specifies perfect negative correlation. All of the data points would fall on a straight line slanted downward.

3. r = +1 specifies perfect positive correlation. All of the data points would fall on a straight line slanted upward.

4. $r \approx 0$ indicates little or no consistent linear pattern or a horizontal pattern.

3.25 Coefficient values near zero indicate that there is very little or no linear correlation.

<u>Calculating r - the linear correlation coefficient</u>

Preliminary Calculations:

1. Set up a table with the column headings: x, y, x^2, xy and y^2.

2. Insert the bivariate data into corresponding x and y columns. Perform the various algebraic functions to fill in the remaining columns.

3. Sum all columns, that is, find Σx, Σy, Σx^2, Σxy, Σy^2.

4. Double check calculations and summations.

5. Calculate: SS(x) - the sum of squares of x
        SS(y) - the sum of squares of y
        SS(xy) - the sum of squares of xy
   where:
        $SS(x) = \Sigma x^2 - ((\Sigma x)^2/n)$
        $SS(y) = \Sigma y^2 - ((\Sigma y)^2/n)$
        $SS(xy) = \Sigma xy - ((\Sigma x \cdot \Sigma y)/n)$ ...

```
Final Calculation:

6. Calculate r:
```

$$r = \frac{SS(xy)}{\sqrt{SS(x)\,SS(y)}}$$

(round to the nearest hundredth)

7. Retain the <u>summations</u> and the <u>sums of squares</u>, as they will be needed for later calculations.

NOTE: Remember $SS(x) \neq \sum x^2$, $SS(y) \neq \sum y^2$ and $SS(xy) \neq \sum xy$.

OR:

The MINITAB command to calculate the correlation coefficient can be found in JES2-p128. MINITAB commands to complete the preliminary calculations (summations and sums of squares) can be found in JES2-pp128&129.

For the calculation of 'r', x and y can be assigned to either variable. In exercise 3.27, let x = age and y = score.

3.27

x	y	x^2	xy	y^2
8	31	64	248	961
9	25	81	225	625
9	40	81	360	1600
10	27	100	270	729
11	35	121	385	1225
9	29	81	261	841
9	25	81	225	625
9	34	81	306	1156
11	27	121	297	729
11	36	121	396	1296
96	309	932	2973	9787

a. $SS(x) = \sum x^2 - ((\sum x)^2/n) = 932 - (96^2/10) = \underline{10.4}$

b. $SS(y) = \sum y^2 - ((\sum y)^2/n) = 9787 - (309^2/10) = \underline{238.9}$

c. $SS(xy) = \sum xy - ((\sum x \cdot \sum y)/n) = 2973 - (96 \cdot 309/10) = \underline{6.6}$

d. $r = SS(xy)/\sqrt{SS(x) \cdot SS(y)} = 6.6/\sqrt{10.4 \cdot 238.9} = 0.132 = \underline{0.13}$

```
┌─────────────────────────────────────────────────────────────────────┐
│          Estimating r - the linear correlation coefficient          │
│ 1. Draw as small a rectangle as possible that encompasses all of    │
│    the data on the scatter diagram. (Diagram should cover a "square │
│    window" - same length and width)                                 │
│ 2. Measure the width.                                               │
│ 3. Let k = the number of times the width fits along the length      │
│              or in other words: length/width.                       │
│                 1                                                    │
│ 4. r ≈ ±(1 -  ───)                                                   │
│                 k                                                    │
│ 5. Use +, if the rectangle is slanted positively or upward.         │
│    Use -, if the rectangle is slanted negatively or downward.       │
└─────────────────────────────────────────────────────────────────────┘
```

3.29 a. Estimate r to be near 2/3 or 0.7.

b.

Data	x	y	x^2	xy	y^2
1	2	5	4	10	25
2	3	5	9	15	25
3	3	7	9	21	49
4	4	5	16	20	25
5	4	7	16	28	49
6	5	7	25	35	49
7	5	8	25	40	64
8	6	6	36	36	36
9	6	9	36	54	81
10	6	8	36	48	64
11	7	7	49	49	49
12	7	9	49	63	81
13	7	10	49	70	100
14	8	8	64	64	64
15	8	9	64	72	81
Σ	81	110	487	625	842

$$SS(x) = \Sigma x^2 - ((\Sigma x)^2/n) = 487 - (81^2/15) = 49.6$$
$$SS(y) = \Sigma y^2 - ((\Sigma y)^2/n) = 842 - (110^2/15) = 35.333$$
$$SS(xy) = \Sigma xy - ((\Sigma x \cdot \Sigma y)/n) = 625 - (81 \cdot 110/15) = 31.0$$

$$r = SS(xy)/\sqrt{SS(x) \cdot SS(y)} = 31.0/\sqrt{49.6 \cdot 35.333} = \underline{0.741} = \underline{0.74}$$

```
┌─────────────────────────────────────────────────────────────────────┐
│ * What is the r value of 0.74 in exercise 3.29 telling you?         │
│ How do your estimated r and calculated r compare?  They should be   │
│ relatively close. (See the bottom of the next page for answer.)     │
└─────────────────────────────────────────────────────────────────────┘
```

3.31

x	y	x²	xy	y²
2.5	40	6.25	100.0	1600
3.0	43	9.00	129.0	1849
4.0	30	16.00	120.0	900
3.5	35	12.25	122.5	1225
2.7	42	7.29	113.4	1764
4.5	19	20.25	85.5	361
3.8	32	14.44	121.6	1024
2.9	39	8.41	113.1	1521
5.0	15	25.00	75.0	225
2.2	44	4.84	96.8	1936
34.1	339	123.73	1076.9	12,405

a. $SS(x) = \sum x^2 - ((\sum x)^2/n) = 123.73 - (34.1^2/10) = \underline{7.449}$

b. $SS(y) = \sum y^2 - ((\sum y)^2/n) = 12,405 - (339^2/10) = \underline{912.9}$

c. $SS(xy) = \sum xy - ((\sum x \cdot \sum y)/n) = 1076.9 - (34.1 \cdot 339/10)$
$$= \underline{-79.09}$$

d. $r = SS(xy)/\sqrt{SS(x) \cdot SS(y)} = -79.09/\sqrt{7.449 \cdot 912.9} = \underline{-0.96}$

** What is the r value of -0.96 in exercise 3.31 telling you? (See the bottom of this page for answer.)

3.33 a. Summations from extensions table: $n = 7$, $\sum x = 161$, $\sum y = 201$, $\sum x^2 = 4065$, $\sum xy = 4321$, $\sum y^2 = 6275$

$SS(x) = \sum x^2 - ((\sum x)^2/n) = 4065 - (161^2/7) = 362.0$

$SS(y) = \sum y^2 - ((\sum y)^2/n) = 6275 - (201^2/7) = 503.429$

$SS(xy) = \sum xy - ((\sum x \cdot \sum y)/n) = 4321 - (161 \cdot 201/7) = -302.0$

$r = SS(xy)/\sqrt{SS(x) \cdot SS(y)} = -302.0/\sqrt{362 \cdot 503.429} = \underline{-0.707}$

---

*(3.29) There is a positive relationship between the variables, that is, as x increases, y increases. In this case, as the hours studied increased, the grade on the exam increased. What can one deduce from this?

**(3.31) The negative indicates that as x's increase, y's decrease. The -0.96 is close to -1, also indicating a strong negative linear relationship. In this case, as the weight of the automobile increases, the gas mileage decreases. That is, the heavier the car, the lower the gas mileage.

---

b. As the level of work satisfaction decreased, the inclination to leave a job increased. If a person does not enjoy their job, they are more likely to be looking for another job.

SECTION 3.3 MARGIN EXERCISES

3.35 a. $\Sigma x^2 = 13,717$, SS(x) = 1396.9
      $\Sigma y^2 = 15,298$, SS(y) = 858.0
      $\Sigma xy = 14,257$, SS(xy) = 919.0

b. Summations (Σ's) are sums of data values. Sum of squares (SS( )) are parts of complex formulas; they are calculated separately as preliminary values.

3.36 a. $\hat{y} = 14.9 + 0.66(20) = \underline{28.1}$

   $\hat{y} = 14.9 + 0.66(50) = \underline{47.9}$

b. Yes, the line of best fit is made up of all points that satisfy its equation.

3.37 a. Summations from extensions table: n = 5, $\Sigma x = 14$,
      $\Sigma y = 380$, $\Sigma x^2 = 50$, $\Sigma xy = 1110$, $\Sigma y^2 = 29,400$

   SS(x) = $\Sigma x^2 - ((\Sigma x)^2/n)$ = 50 - (14²/5) = 10.8
   SS(xy) = $\Sigma xy - ((\Sigma x \cdot \Sigma y)/n)$ = 1110 - (14·380/5) = 46

   b_1 = SS(xy)/SS(x) = 46/10.8 = 4.259

   b_0 = $[\Sigma y - b_1 \cdot \Sigma x]/n$ = [380 - (4.259·14)]/5 = 64.0748

   $\underline{\hat{y} = 64.1 + 4.26x}$

b. At x = 1, $\hat{y}$ = 64.1 + 4.26(1) = 68.36; thus (1,68.4)
   At x = 3, $\hat{y}$ = 64.1 + 4.26(3) = 76.9; thus (3,76.9)
   Points (1,68.4) and (3,76.9) are used to locate the line.

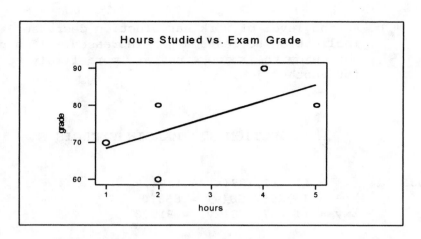

c. Yes, as the hours studied increased, the exam grades appear to increase, also.

3.38 a. $\hat{y} = 14.9 + 0.66(40) = 41.3 = \underline{41}$
b. No
c. 41 is the average number of sit-ups expected for students who do 40 push-ups.

3.39 a.

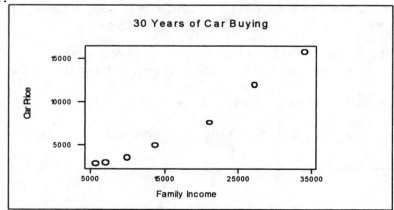

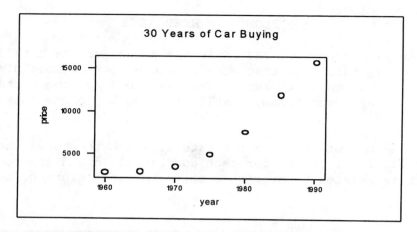

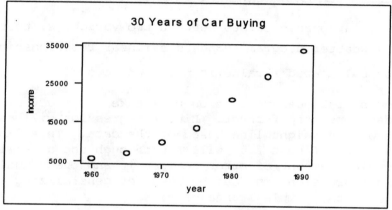

b. The patterns are definitely elongated, however they do not appear to be exactly linear.

c. Yes, dividing median incomes by 2 comes fairly close to the average price of a new car.

3.40 The vertical scale shown on figure 3-24 is located at x = 58 and therefore is not the y-axis; the y = 80 occurs at x = 58. Remember, the x-axis is the vertical line located at x = 0.

3.41 (61, 95) (67, 130)

$$b_1 \approx \frac{y_2 - y_1}{x_2 - x_1} = \frac{130 - 95}{67 - 61} = \frac{35}{6} = 5.8$$

$$b_0 \approx y - b_1 x = 130 - 5.83(67) = -260.61$$

-- 67 --

3.43 a. The y-intercept of $23.65 is the amount of the total monthly telephone cost when x, the number of long distance calls, is equal to zero. That is, when no long distance calls are made there is still the monthly phone charge of $23.65.

b. The slope of $1.28 is the rate at which the total phone bill will increase for each additional long distance call; it is related to average cost of the long distance calls.

If a linear relationship exists between two variables, that is,

1. its scatter diagram suggests a linear relationship

2. its calculated r value is not near zero

the techniques of linear regression will take the study of bivariate data one step further. Linear regression will calculate an equation of a straight line based on the data. This line, also known as the line of best fit, will fit through the data with the smallest possible amount of error between it and the actual data points. The regression line can be used for generalizing and predicting over the sampled range of x.

FORM OF A LINEAR REGRESSION LINE

$$\hat{y} = b_0 + b_1 x$$

where $\hat{y}$ (y hat) = predicted y
b_0 (b sub zero) = y intercept
b_1 (b sub one) = slope of the line
x = independent data value.

3.45 $\hat{y}$ = 7.31 - 0.01x when x = 50 is
$\hat{y}$ = 7.31 - 0.01(50) = 6.81

The predicted value is 6.81(10,000) or $68,100

3.47 $\hat{y}$ = 1.85 + 0.30x, when x = 3.0
$\hat{y}$ = 1.85 + 0.30(3.0) = 2.75

CALCULATING $\hat{y} = b_0 + b_1 x$ - THE EQUATION OF THE LINE OF BEST FIT

1. Retrieve preliminary calculations from previous r calculations.

2. Calculate b_1 where $b_1 = \dfrac{SS(xy)}{SS(x)}$

3. Calculate b_0 where $b_0 = \dfrac{1}{n}(\Sigma y - b_1 \Sigma x)$

$\hat{y}$ = predicted value of y (based on the regression line)

NOTE: See Review Lessons for additional information about the concepts of slope and intercept of a straight line.

DRAWING THE LINE OF BEST FIT ON THE SCATTER DIAGRAM

1. Pick two x-values that are <u>within</u> the interval of the data x-values. (one value near either end of the domain)

2. Substitute these values into the calculated $\hat{y} = b_0 + b_1 x$ equation and find the corresponding $\hat{y}$ values.

3. Plot these points on the scatter diagram in such a manner that they are distinguishable from the actual data points.

4. Draw a straight line connecting these two points. This line is a graph of the line of best fit.

5. Plot a third point, the ordered pair $(\overline{x}, \overline{y})$ as an additional check. It should be a point on the line of best fit.

OR:

MINITAB commands to find the equation of the line of best fit and also draw it on a scatter diagram can be found in JES2-pp141&142. **NOTE:** The **REGRess** command takes the y values first. You need not adjust the way you input your data; just watch the order of your data and the order of the command.

3.49 a.

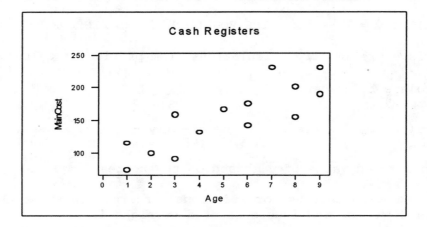

b. Summations from extensions table: n = 14, $\sum x = 72$,
$\sum y = 2163$, $\sum x^2 = 476$, $\sum xy = 12,677$, $\sum y^2 = 366,311$

$SS(x) = \sum x^2 - ((\sum x)^2/n) = 476 - (72^2/14) = 105.71$
$SS(xy) = \sum xy - ((\sum x \cdot \sum y)/n) = 12,677 - (72 \cdot 2163/14) = 1553.00$

$b_1 = SS(xy)/SS(x) = 1553.00/105.71 = 14.691$

$b_0 = [\sum y - b_1 \cdot \sum x]/n = [2163 - (14.691 \cdot 72)]/14 = 78.95$

$\hat{y} = \underline{78.95 + 14.69x}$

c. $\hat{y} = 78.95 + 14.69x$, when x = 8
$\hat{y} = 78.95 + 14.69(8) = \underline{196.47}$

d. \$196.47 can be interpreted as the expected average cost of
maintenance for all such 8-year-old cash registers.

3.51 Summations from extensions table: n = 10, $\sum x = 325$,
$\sum y = 349$, $\sum x^2 = 11,107$, $\sum xy = 11,608$, $\sum y^2 = 12,889$

$SS(x) = \sum x^2 - ((\sum x)^2/n) = 11,107 - (325^2/10) = 544.5$
$SS(y) = \sum y^2 - ((\sum y)^2/n) = 12,889 - (349^2/10) = 708.9$
$SS(xy) = \sum xy - ((\sum x \cdot \sum y)/n) = 11,608 - (325 \cdot 349/10) = 265.5$

a. $r = SS(xy)/\sqrt{SS(x) \cdot SS(y)} = 265.5/\sqrt{544.5 \cdot 708.9} = \underline{0.427}$

b. $b_1 = SS(xy)/SS(x) = 265.5/544.5 = 0.487603$

$b_0 = [\sum y - b_1 \cdot \sum x]/n = [349 - (0.487603 \cdot 325)]/10 = 19.05289$

$\hat{y} = \underline{19.1 + 0.488x}$

Exercises 3.53-3.56

To find percentages based on the grand total
- divide each count by the grand total

To find percentages based on the row totals
- divide each count by its corresponding row total
- each row should add up to 100%

To find percentages based on the column totals
- divide each count by its corresponding column total
- each column should add up to 100%

3.53 a.

	Parents	Relatives	Friends	Combination	Others	Total
Rural	75	46	24	37	2	184
Non	30	32	44	57	7	170
Total	105	78	68	94	9	354

b.

	Parents	Relatives	Friends	Combination	Others	Total
Rural	21.2%	13.0%	6.8%	10.5%	0.6%	52.1%
Non	8.5%	9.0%	12.4%	16.1%	2.0%	48.0%
Total	29.7%	22.0%	19.2%	26.6%	2.6%	100.1%*

*The 100.1% is due to round-off error.

c.

	Parents	Relatives	Friends	Combination	Others	Total
Rural	71.4%	59.0%	35.3%	39.4%	22.2%	52.0%
Non	28.6%	41.0%	64.7%	60.6%	77.8%	48.0%
Total	100%	100%	100%	100%	100%	100%

d.	Parents	Relatives	Friends	Combination	Others	Total
Rural	40.8%	25.0%	13.0%	20.1%	1.1%	100%
Non	17.6%	18.8%	25.9%	33.5%	4.1%	100%
Total	29.7%	22.0%	19.2%	26.6%	2.5%	100%

e.

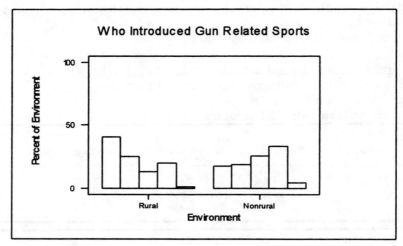

3.55 a.

	Less than 6mo	6 mo - 1 yr	More than 1yr	Total
Under 28	413	192	295	900
28 -40	574	208	218	1000
Over 40	653	288	259	1200
Total	1640	688	772	3100

b.

	Less than 6mo	6 mo - 1 yr	More than 1yr	Total
Under 28	13.3%	6.2%	9.5%	29.0%
28 -40	18.5%	6.7%	7.0%	32.2%
Over 40	21.1%	9.3%	8.4%	38.8%
Total	52.9%	22.2%	24.9%	100%

c.

	Less than 6mo	6 mo - 1 yr	More than 1yr	Total
Under 28	45.9%	21.3%	32.8%	100%
28 -40	57.4%	20.8%	21.8%	100%
Over 40	54.4%	24.0%	21.6%	100%
Total	52.9%	22.2%	24.9%	100%

d.

	Less than 6mo	6 mo - 1 yr	More than 1yr	Total
Under 28	25.2%	27.9%	38.2%	29.0%
28 -40	35.0%	30.2%	28.2%	32.3%
Over 40	39.8%	41.9%	33.6%	38.7%
Total	100%	100%	100%	100%

e.

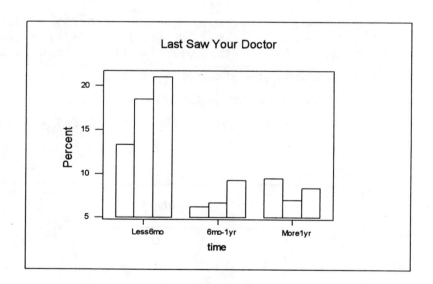

3.57 a. correlation   b. regression   c. correlation
      d. regression   e. correlation

3.59

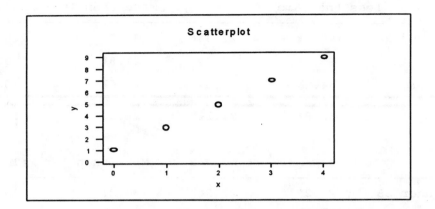

*What would you expect for the correlation coefficient if <u>all</u> of these points fall exactly on a straight line? (See the bottom of the next page for answer.)

Summations from extensions table: $n = 5$, $\sum x = 10$, $\sum y = 25$, $\sum x^2 = 30$, $\sum xy = 70$, $\sum y^2 = 165$

$SS(x) = \sum x^2 - ((\sum x)^2/n) = 30 - (10^2/5) = 10.0$
$SS(y) = \sum y^2 - ((\sum y)^2/n) = 165 - (25^2/5) = 40.0$
$SS(xy) = \sum xy - ((\sum x \cdot \sum y)/n) = 70 - (10 \cdot 25/5) = 20.0$

$r = SS(xy)/\sqrt{SS(x) \cdot SS(y)} = 20.0/\sqrt{10.0 \cdot 40.0} = \underline{1.00}$

$b_1 = SS(xy)/SS(x) = 20.0/10.0 = 2.0$

$b_0 = [\sum y - b_1 \cdot \sum x]/n = [25 - (2.0 \cdot 10)]/5 = 1.0$

$\underline{\hat{y} = 1.0 + 2.0x}$

3.61 a.

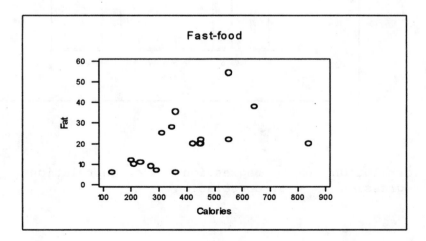

-- 74 --

Summations from extensions table: n = 18, Σx = 7055,
Σy = 367, Σx^2 = 3,287,783, Σxy = 164,625, Σy^2 = 10,309

SS(x) = Σx^2 - (($\Sigma x)^2$/n) = 3,287,783 - (7055^2/18)
$\qquad\qquad\qquad$ = 522,614.944
SS(y) = Σy^2 - (($\Sigma y)^2$/n) = 10,309 - (367^2/18) = 2826.2777
SS(xy) = Σxy - (($\Sigma x \cdot \Sigma y)$/n) = 164,625 - (7055·367/18)
$\qquad\qquad\qquad$ = 20,781.3889

b. r = SS(xy)/$\sqrt{SS(x) \cdot SS(y)}$

$\qquad$ = 20,781.3889/$\sqrt{522614.944 \cdot 2826.2777}$ = <u>0.541</u>

c. b_1 = SS(xy)/SS(x) = 20,781.3889/522,614.944 = 0.039764

$\qquad b_0$ = [Σy - $b_1 \cdot \Sigma x$]/n = [367 - (0.039764·7055)]/18 = 4.80351

$\qquad \hat{y}$ <u>= 4.80 + 0.04x</u>

d. There is slight correlation between fast-food calories and
$\qquad$ the corresponding amount of fat. Generally, if the
$\qquad$ calories increase so does the fat content.

3.63

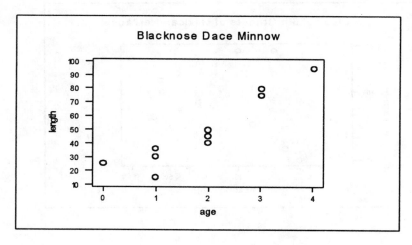

*(3.59) If all of the points fall exactly on a straight line,
perfect positive or negative correlation has occurred. The r value
will be either +1 or -1, depending on the upward or downward trend
of the data points.

Summations from extensions table: n = 10, Σx = 19, Σy = 491, Σx^2 = 49, Σxy = 1196, Σy^2 = 30,221

$SS(x) = \Sigma x^2 - ((\Sigma x)^2/n) = 49 - (19^2/10) = 12.9$
$SS(y) = \Sigma y^2 - ((\Sigma y)^2/n) = 30,221 - (491^2/10) = 6112.9$
$SS(xy) = \Sigma xy - ((\Sigma x \cdot \Sigma y)/n) = 1196 - (19 \cdot 491/10) = 263.1$

b. $r = SS(xy)/\sqrt{SS(x) \cdot SS(y)}$

   $= 263.1/\sqrt{12.9 \cdot 6112.9} = \underline{0.937}$

c. $b_1 = SS(xy)/SS(x) = 263.1/12.9 = 20.40$

   $b_0 = [\Sigma y - b_1 \cdot \Sigma x]/n = [491 - (20.40 \cdot 19)]/10 = 10.34$

   $\underline{\hat{y} = 10.34 + 20.40x}$

d. There is a strong correlation between the age of a blacknose dace minnow and its length. Generally, if the age increases so does the length.

3.65 a.

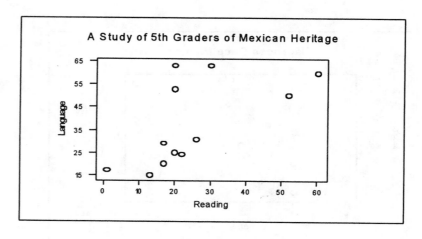

b.

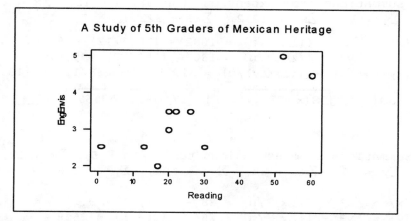

c.

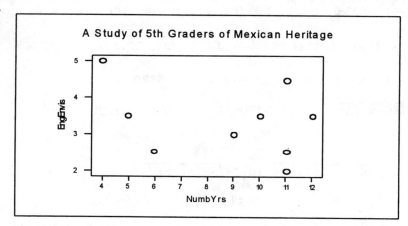

d. Summations from extensions table: n = 12, Σx = 298,
   Σy = 450, Σx^2 = 10,312, Σxy = 13,289, Σy^2 = 20,764

$SS(x) = \Sigma x^2 - ((\Sigma x)^2/n) = 10,312 - (298^2/12) = 2911.67$
$SS(y) = \Sigma y^2 - ((\Sigma y)^2/n) = 20,764 - (450^2/12) = 3889.00$
$SS(xy) = \Sigma xy - ((\Sigma x \cdot \Sigma y)/n) = 13,289 - (298 \cdot 450/12) = 2114.00$

$r = SS(xy)/\sqrt{SS(x) \cdot SS(y)} = 2114.0/\sqrt{2911.67 \cdot 3889.0} = \underline{0.63}$

e. Summations from extensions table: n = 12, Σx = 298,
   Σy = 38, Σx^2 = 10,312, Σxy = 1076, Σy^2 = 130

$SS(x) = \Sigma x^2 - ((\Sigma x)^2/n) = 10,312 - (298^2/12) = 2911.67$
$SS(y) = \Sigma y^2 - ((\Sigma y)^2/n) = 130 - (38^2/12) = 9.67$
$SS(xy) = \Sigma xy - ((\Sigma x \cdot \Sigma y)/n) = 1076 - (298 \cdot 38/12) = 132.33$

$r = SS(xy)/\sqrt{SS(x) \cdot SS(y)} = 132.33/\sqrt{2911.67 \cdot 9.67} = \underline{0.79}$

f. Summations from extensions table: $n = 12$, $\sum x = 106$, $\sum y = 38$, $\sum x^2 = 1032$, $\sum xy = 322.5$, $\sum y^2 = 130$

$SS(x) = \sum x^2 - ((\sum x)^2/n) = 1032 - (106^2/12) = 95.67$
$SS(y) = \sum y^2 - ((\sum y)^2/n) = 130 - (38^2/12) = 9.67$
$SS(xy) = \sum xy - ((\sum x \cdot \sum y)/n) = 322.5 - (106 \cdot 38/12) = -13.17$

$r = SS(xy)/\sqrt{SS(x) \cdot SS(y)} = -13.17/\sqrt{95.67 \cdot 9.67} = \underline{-0.43}$

3.67 a. Summations from extensions table: $n = 5$, $\sum x = 12$, $\sum y = 36$, $\sum x^2 = 38$, $\sum xy = 101$, $\sum y^2 = 286$

$SS(x) = \sum x^2 - ((\sum x)^2/n) = 38 - (12^2/5) = 9.2$
$SS(y) = \sum y^2 - ((\sum y)^2/n) = 286 - (36^2/5) = 26.8$
$SS(xy) = \sum xy - ((\sum x \cdot \sum y)/n) = 101 - (12 \cdot 36/5) = 14.6$

$r = SS(xy)/\sqrt{SS(x) \cdot SS(y)} = 14.6/\sqrt{9.2 \cdot 26.8} = \underline{0.9298}$

$b_1 = SS(xy)/SS(x) = 14.6/9.2 = 1.5870$

$b_1 \sqrt{SS(x)/SS(y)} = 1.5870 \sqrt{9.2/26.8} = \underline{0.9298}$

b. $b_1 \sqrt{SS(x)/SS(y)} = [b_1] \sqrt{SS(x)/SS(y)}$

$$= [SS(xy)/SS(x)] \cdot [\sqrt{SS(x)/SS(y)}]$$

$$= [SS(xy) \cdot 1/SS(x)] \cdot [\sqrt{SS(x)} \cdot \sqrt{1/SS(y)}]$$

$$= SS(xy) \cdot [1/SS(x)] \cdot [\sqrt{SS(x)}] \cdot [\sqrt{1/SS(y)}]$$

$$= SS(xy) \cdot [1/\sqrt{SS(x)}] \cdot [1/\sqrt{SS(y)}]$$

$$= SS(xy) \cdot [1/\sqrt{SS(x) \cdot SS(y)}]$$

$$= SS(xy)/\sqrt{SS(x) \cdot SS(y)}$$

$$= r$$

CHAPTER 4 ∇ PROBABILITY

Chapter Preview

Chapter 4 deals with the basic theory and concepts of probability. Probability, in combination with the descriptive techniques in the previous chapters, allows us to proceed into inferential statistics in later chapters.

SECTION 4.1 EXERCISES

4.1 Each student will get different results; the two answers will each be a fraction with a denominator of 10. If your results are 6 heads and 4 tails:
a. Relative frequency of H is <u>0.6</u>
b. Relative frequency of T is <u>0.4</u>

4.3 Each student will get different results. The four answers will each be a fraction with a denominator of 25.

MINITAB commands to generate random integers and tally the findings can be found in JES2-p162.

A variation is necessary for exercise 4.5a.   Use: RANDom 50 C1;
INTEger 1 6.
TALLy C1;
COUNts;
PERCents.

A variation is necessary for exercise 4.5b.   Use: RANDom 100 C2;
INTEger 0 1.
TALLy C2;
COUNts;
PERCents.

4.5 Note: Each will get different results. MINITAB results on one run were:
a. Relative frequency for: 1 - <u>0.22,</u> 2 - <u>0.16,</u> 3 - <u>0.14,</u> 4 - <u>0.22,</u> 5 - <u>0.16,</u> 6 - <u>0.10</u>

b. Relative frequency for: H - <u>0.58,</u> T - <u>0.42</u>

4.7   P'(5) = 9/40 = <u>0.225</u>

4.8   All three are calculated by dividing the experimental count
by the sample size.

SECTION 4.2 EXERCISES

4.9 You can expect a 1 to occur approximately 1/6th of the time
when you roll a die repeatedly.

4.11 a. 36,000,000/200,000,000 = <u>0.18</u>

b. 19,000/200,000,000 = <u>0.000095</u>

4.13 Each student will get different results.   These are the
results I obtained: [Note: 12 is an ordered pair (1,2)]

12	65	15	32	54	12	52	63	64	62
66	44	42	45	42	35	54	66	54	32
31	12	23	33	26	33	32	23	46	64
63	63	35	54	52	55	56	26	11	44
11	61	46	11	45	55	15	33	43	11

a. P'(white die is odd) = 27/50 = <u>0.54</u>

b. P'(sum is 6) = 7/50 = <u>0.14</u>

c. P'(both dice show odd number) = 14/50 = <u>0.28</u>

d. P'(number on color die is larger) = 16/50 = <u>0.32</u>

4.15 Each student will get different results.   These are the
results I obtained:

n(heads)/10	P'(head)/set of 10	Cum.P'(head)
6	0.6	6/10 = 0.60
3	0.3	9/20 = 0.45
5	0.5	14/30 = 0.47
5	0.5	19/40 = 0.48
7	0.7	26/50 = 0.52

4	0.4	30/60 = 0.50
6	0.6	36/70 = 0.51
6	0.6	42/80 = 0.52
6	0.6	48/90 = 0.53
5	0.5	53/100 = 0.53
3	0.3	56/110 = 0.51
4	0.4	60/120 = 0.50
7	0.7	67/130 = 0.52
3	0.3	70/140 = 0.50
6	0.6	76/150 = 0.51
3	0.3	79/160 = 0.49
7	0.7	86/170 = 0.51
7	0.7	93/180 = 0.52
4	0.4	97/190 = 0.51
6	0.6	103/200 = 0.52

Observed Probability of Heads in Sets of Ten

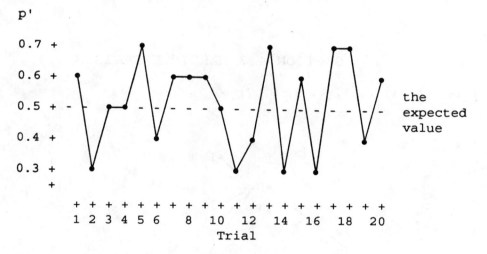

The observed probability varies above and below 0.5, but seems to average approximately 0.5.

Cumulative Observed Probability of Heads from Sets of Ten

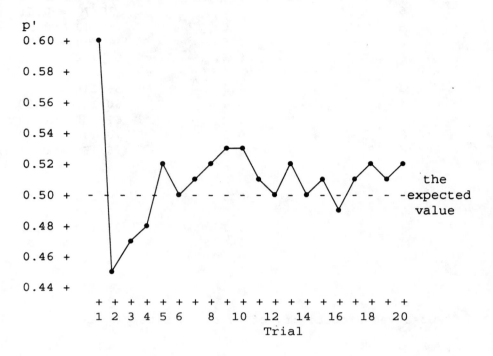

p'
0.60 +
0.58 +
0.56 +
0.54 +
0.52 +
0.50 + - - - - - - - - - - - - - - - -expected
0.48 +   value
0.46 +
0.44 +
    +
    1 2 3 4 5 6   8 10   12 14   16 18   20
                        Trial

SECTION 4.3 MARGIN EXERCISES

4.16 {0, 1, 2, 3, 4, 5, 6, 7, 8, 9}

4.17

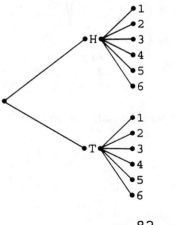

4.18   a.   S = {$1, $5, $10, $20}

b.

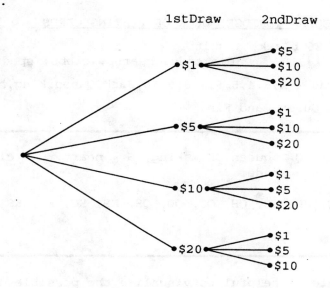

1stDraw   2ndDraw

$1 → $5, $10, $20

$5 → $1, $10, $20

$10 → $1, $5, $20

$20 → $1, $5, $10

c.

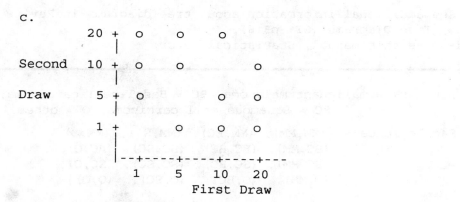

A REGULAR (BRIDGE) DECK OF PLAYING CARDS

52 cards: 26 red, 26 black, 4 suits

(diamonds, hearts, clubs, spades)

each suit - 13 cards - 2,3,4,5,6,7,8,9,10,Jack,Queen,King,Ace

face cards = Jack, Queen, and King

4.19 Let J = jack, Q = queen, K = king, H = heart, C = club,
D = diamond, S = spade.

S = {JH, JC, JD, JS, QH, QC, QD, QS, KH, KC, KD, KS}

A tree diagram would be helpful to visualize the possible outcomes.
Follow the branches for a list of outcomes.

NOTE: See additional information about tree diagrams in Review
Lessons, Tree Diagrams (ST*-p316).
*(ST denotes this manual, Statistical Tutor)

4.21 Let: MM = malignant melanoma, BC = Basal-cell carcinoma,
SC = Squamous-cell carcinoma, O = other.

Sample space = {(MM,MM), (MM,BC), (MM,SC), (MM,O),
(BC,MM), (BC,BC), (BC,SC), (BC,O),
(SC,MM), (SC,BC), (SC,SC), (SC,O),
(O,MM), (O,BC), (O,SC), (O,O)}

A tree diagram is helpful for exercise 4.23.

4.23 S = {H1, H2, H3, H4, H5, H6, T1, T2, T3, T4, T5, T6}

4.25 a. S = {HH, HT, T1, T2, T3, T4, T5, T6}

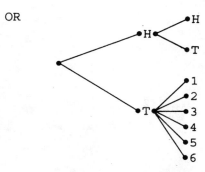

OR

b. <u>0.5</u>, the same as the probability of a tail on first trial.

TABLe C1 C2 divides the data based on the possible values in each column. The results (count) are in tabular form, giving row and column totals.

NOTE: P' is the notation for an experimental or empirical probability.

4.27 Everyone's results will be different. One example of observed probabilities are:

a. P'(H) = 90/200 = <u>0.45</u>

b. P'(3) = 30/200 = <u>0.15</u>

c. P'(H,3) = 12/200 = <u>0.06</u>

SECTION 4.4 MARGIN EXERCISES

4.29 P(5) = 4/36;   P(6) = 5/36;   P(7) = 6/36;   P(8) = 5/36;
P(9) = 4/36;   P(10) = 3/36;   P(11) = 2/36; P(12) = 1/36

4.30 $P(P) = 4 \cdot P(F)$
$P(P) + P(F) = 1$
$4 \cdot P(F) + P(F) = 1$
$5 \cdot P(F) = 1$
$P(F) = 1/5$
$P(P) = 4 \cdot P(F) = 4 \cdot (1/5) = \underline{4/5}$

4.31 a. $\dfrac{2}{2 + 7} = \dfrac{2}{9}$ b. 7:2

4.32 a. $\dfrac{64}{150,000} = \dfrac{1}{2343.75} = 0.0004267$

   b. (senior:nonsenior) 2400:1400 (3800-2400=1400)
         reduces to $\underline{12:7}$

   c. (no play:play) 2336:64 (2400-64=2336)
         reduces to $\underline{37:1}$ (rounded)

4.33 A = face card

   $P(\overline{A}) = 1 - P(A) = 1 - \dfrac{12}{52} = \dfrac{40}{52}$

SECTION 4.4 EXERCISES

4.35 a. 1/6   b. 3/6   c. 4/6   d. 3/6

The sum of all the probabilities over a sample space is equal to one. $\qquad \sum P(x) = 1$

The probability of event A plus the probability event B plus the probability of event C should be equal to the probability of the sample space, which is equal to one.
$$P(A) + P(B) + P(C) = 1$$

4.37 Let $\quad P(A) = X, \quad P(B) = 2X, \quad P(C) = 4X$

$\qquad X + 2X + 4X = 1$
$\qquad 7X = 1$
$\qquad X = 1/7$

therefore: $\quad P(A) = \underline{1/7,} \quad P(B) = \underline{2/7,} \quad P(C) = \underline{4/7.}$

PROPERTIES OF PROBABILITY

1. $0 \leq P(A) \leq 1$ The probability of an event must be a value between 0 and 1, inclusive.

2. $\sum P(A) = 1$ The sum of all the probabilities for each event in the sample space equals 1.

4.39 The three success ratings (highly successful, successful, and not successful) appear to be non intersecting, and their union appears to be the entire sample space. If this is true, none of the three sets of probabilities are appropriate.

Judge A has a total probability of 1.2. The total must be exactly 1.0.

Judge B has a negative probability of -0.1 for one of the events. All probability numbers are between 0.0 and 1.0.

Judge C has a total probability of 0.9. The total must be exactly 1.0.

In exercise 4.41, add the rows and columns first, to find marginal totals.

4.41 a. (35+20)/100 = 0.55

b. (20+20)/100 = 0.40

4.43 a. 1:4   b. 1/750,001   c. 20/21

COMPLEMENT - Probability of A complement = $P(\overline{A})$

$$P(\overline{A}) = P(\text{not } A) = 1 - P(A)$$

4.45 P(female) = 1/9; P(male) = 1 - (1/9) = 8/9 = 0.89

SECTION 4.5 MARGIN EXERCISES

4.47 a. Yes, they can not occur at the same time; i.e., a student can not be both male and female.
b. No, they can occur at the same time; i.e., a student can be both male and registered for statistics.
c. No, they can occur at the same time; i.e., a student can be both female and registered for statistics.
d. Yes, the probability of being female at this college plus the probability of being male at this college equals one.
e. No, the two events do not include all of the students.
f. Yes, in both situations there are no common elements shared by the two events.
g. No, two complementary events comprise the sample sample; two mutually exclusive events do not necessarily make up the whole sample space.

4.48 a. A & C and A & E are mutually exclusive because they cannot occur at the same time.
b. P(A or C) = P(A) + P(C) = 6/36 + 6/36 = 12/36
P(A or E) = P(A) + P(E) = 6/36 + 5/36 = 11/36
P(C or E) = P(C) + P(E) - P(C and E)
         = 6/36 + 5/36 - 1/36 = 10/36

> **Mutually Exclusive Events** - events that cannot occur at the same time (they have no sample points in common).
>
> **Not Mutually Exclusive Events** - events that can occur at the same time (they have sample points in common).

4.49 a. **Not mutually exclusive.** *One head* belongs to both events, therefore the two event intersect.

   b. **Not mutually exclusive.** All sales that *exceed $1000* also *exceed $100*, therefore $1200 belongs to both events and the two events have an intersection.

   c. **Not mutually exclusive.** The student selected could be both *male* and *over 21*, therefore the two events have an intersection.

   d. **Mutually exclusive.** The total cannot be both *less than 7* and *more than 9* at the same time, therefore there is no intersection between these two events.

PROBABILITY - THE ADDITION RULE

$P(A \text{ or } B) = P(A) + P(B) - P(A \text{ and } B)$ if A and B are **not** mutually exclusive

The **probability of event A or event B** is equal to the probability of event A plus the probability of event B, minus the probability of events A and B occurring at the same time (otherwise that common probability is counted twice).

$P(A \text{ or } B) = P(A) + P(B)$ if A and B are mutually exclusive

The **probability of event A or event B** is equal to the probability of event A plus the probability of event B, if A and B have nothing in common (i.e., they cannot occur at the same time).

$P(A \text{ and } B) = 0$ if A and B are mutually exclusive

The **probability of events A and B** occurring at the same time is impossible if A and B are mutually exclusive.

4.51 If two events are mutually exclusive, then there is no intersection. The event, *A and B*, is the intersection. If no intersection, then P(A and B) = 0.0.

4.53 a. $P(\overline{A})$ = 1 - 0.3 = <u>0.7</u>

   b. $P(\overline{B})$ = 1 - 0.4 = <u>0.6</u>

   c. P(A or B) = 0.3 + 0.4 = <u>0.7</u>

   d. P(A and B) = <u>0.0</u>   (Mutually exclusive events have no intersection.)

4.55 No. *Female* students can be *working* students. Further, if the probabilities are correct, there must be an intersection otherwise the total probability would be more than 1.0.

4.57 a. No, they can occur at the same time; i.e., a patient can be both female and in ICU.
   b. Yes, they cannot occur at the same time; i.e., a patient cannot be in ICU and the Surgical Unit at the same time.
   c. P(ICU or Female) = 20/44 + 15/44 - 9/44 = <u>26/44</u>
   d. P(ICU or Male) = 20/44 + 29/44 - 11/44 = <u>38/44</u>

4.59 Let U = belongs to union and T = makes more than $12

   Given info: P(U) = 0.60,   P(T) = 0.90,   P(U and T) = 0.40

       P(U or T) = P(U) + P(T) - P(U and T)
       P(U or T) = 0.60 + 0.90 - 0.40
       P(U or T) = 1.10

   Since the probability of an event cannot be larger than 1.0, the information must be incorrect.

SECTION 4.6 MARGIN EXERCISES

4.60 a. P(S) = 135/300 = <u>0.45</u>
   b. P(S|viewer was female) = 80/200 = <u>0.40</u>
   c. P(S|viewer was male) = 55/100 = <u>0.55</u>
   d. No, P(S) ≠ P(S|F) ≠ P(S|M)

4.61 P(A and B) = P(A)·P(B) = 0.70 · 0.40 = <u>0.28</u>

4.62 P(R and H) = P(R)·P(H|R) = 0.60 · 0.25 = <u>0.15</u>

4.63 a. P(A and C) = 80/200 = <u>0.40</u>
b. <u>Yes</u>

4.64 a. P(10 matches) =

$$\left(\frac{20}{80}\right)\left(\frac{19}{79}\right)\left(\frac{18}{78}\right)\left(\frac{17}{77}\right)\left(\frac{16}{76}\right)\left(\frac{15}{75}\right)\left(\frac{14}{74}\right)\left(\frac{13}{73}\right)\left(\frac{12}{72}\right)\left(\frac{11}{71}\right) = \frac{6.704425728 \times 10^{11}}{5.974790569 \times 10^{18}}$$

$$= \underline{1:8,911,711}$$

b. P(9 matches) =

$$(10)\left(\frac{20}{80}\right)\left(\frac{19}{79}\right)\left(\frac{18}{78}\right)\left(\frac{17}{77}\right)\left(\frac{16}{76}\right)\left(\frac{15}{75}\right)\left(\frac{14}{74}\right)\left(\frac{13}{73}\right)\left(\frac{12}{72}\right)\left(\frac{60}{71}\right) = \frac{3.656959488 \times 10^{13}}{5.974790569 \times 10^{18}}$$

$$= \underline{1:163,381}$$

c. $\dfrac{1}{8,911,711} + \dfrac{1}{163,381} + \ldots + \dfrac{1}{22} = 0.05870076511$

0.05870076511 = 1/x;   x = <u>17</u>

SECTION 4.6 EXERCISES

<u>Independent Events</u> - when there are independent events, the occurrence of one event **has no effect** on the probability of the other event.

4.65 a. Two events are mutually exclusive if they cannot occur at the same time or they have no elements in common.
b. Two events are independent if the occurrence of one has no effect on the probability of the other.

c. Mutually exclusive has to do with whether or not the events share common elements; while independence has to do with the effect one event has on the other event's probability.

4.67 a. not independent   b. independent

       c. independent   d. not independent

       e. not independent

PROBABILITY - THE MULTIPLICATION RULE & THE CONDITIONAL

The <u>probability of event A given event B</u> has occurred is a conditional probability, written as $P(A|B)$.

For any two events, the <u>probability of events A and B</u> occurring simultaneously is equal to:

1. the probability of event A times the probability of event B, given event A has already occurred: that is:
$$P(A \text{ and } B) = P(A) \cdot P(B|A)$$

OR

2. the probability of event B times the probability of event A, given event B has already occurred: that is:
$$P(A \text{ and } B) = P(B) \cdot P(A|B)$$

For two independent events:

1. the probability of events A and B occurring simultaneously is equal to the probability of event A times the probability of event B
$$P(A \text{ and } B) = P(A) \cdot P(B)$$

2. the conditional probabilities are equal to the single event probabilities
$$P(A|B) = P(A) \quad \text{AND} \quad P(B|A) = P(B)$$

FORMULAS FOR CONDITIONAL PROBABILITIES

$$P(A|B) = \frac{P(A \text{ and } B)}{P(B)} \quad \text{and} \quad P(B|A) = \frac{P(A \text{ and } B)}{P(A)}$$

...

Conditionals can also be computed without the formulas above. Suppose $P(A|B)$ is desired. The word *given*, $(|)$, in the conditional tells what the newly reduced sample space is. The number of elements in the reduced sample space, $n(B)$, becomes the denominator in the probability fraction. The numerator is the number of elements in the reduced sample space that satisfy the first event, $n(A \text{ and } B)$. Therefore: $P(A|B) = \dfrac{n(A \text{ and } B)}{n(B)}$

4.69 a. $P(A \text{ and } B) = P(B) \cdot P(A|B)$
   $0.12 = 0.4 \cdot P(A|B)$; therefore, $P(A|B) = \underline{0.3}$

   b. $P(A \text{ and } B) = P(A) \cdot P(B|A)$
   $0.12 = 0.3 \cdot P(B|A)$; therefore, $P(B|A) = \underline{0.4}$

   c. <u>Yes,</u> A and B are independent.

4.71 1st: $P(A \text{ and } B) = P(B) \cdot P(A|B) = 0.4 \cdot 0.2 = 0.08$

   $P(A \text{ or } B) = P(A) + P(B) - P(A \text{ and } B)$
   $= 0.30 + 0.40 - 0.08 = \underline{0.62}$

4.73 $P(\text{all three Red}) = P(R_1) \cdot P(R_2|R_1) \cdot P(R_3|R_1 \text{ and } R_2)$

   a. with replacement:
   $P(\text{all Red}) = (4/7) \cdot (4/7) \cdot (4/7) = \underline{64/343}$

   b. without replacement:
   $P(\text{all Red}) = (4/7) \cdot (3/6) \cdot (2/5) = 24/210 = \underline{4/35}$

4.75 $P(\text{male victim}) = 1 - 0.62 = 0.38$
   $P(\text{three males}) = (0.38)(0.38)(0.38) = \underline{0.054872}$

4.77 Let C = correct decision, I = incorrect decision

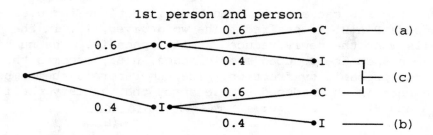

a. P(right decision) = P(C1 and C2) = 0.6·0.6 = <u>0.36</u>

b. P(wrong decision) = P(I1 and I2) = 0.4·0.4 = <u>0.16</u>

c. P(delay) = P[(C1 and I2) or (I1 and C2)]
            = 0.6 · 0.4 + 0.4 · 0.6 = 0.24 + 0.24 = <u>0.48</u>

4.79 Let D = defective, N = nondefective

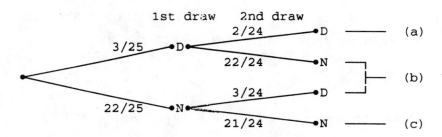

a. P(both defective) = P(D1 and D2) = (3/25)·(2/24) = <u>0.01</u>

b. P(exactly one defective) = P[(D1 and N2) or (N1 and D2)]
              = [(3/25)·(22/24)] + [(22/25)·(3/24)] = <u>0.22</u>

c. P(neither defective) = P(N1 and N2)
                        = (22/25)·(21/24) = <u>0.77</u>

SECTION 4.7 EXERCISES

The probability formula sheet found in ST-pp102&103, may be useful
in completing the exercises.

4.81 P(A or B) = 0.4 + 0.5 - (0.4·0.5) = 0.4 + 0.5 - 0.2 = <u>0.7</u>

4.83 a. P(R and S) = P(R)·P(S) = 0.5·0.3 = <u>0.15</u>

b. P(R or S) = P(R) + P(S) - P(R and S)
= 0.50 + 0.30 - 0.15 = <u>0.65</u>

c. $P(\overline{S})$ = 1 - P(S) = 1 - 0.3 = <u>0.7</u>

d. P(R|S) = P(R and S)/P(S) = 0.15/0.30 = <u>0.5</u>

e. $P(S \text{ and } R) + P(\overline{S} \text{ and } R) = P(R)$;

$0.15 + P(\overline{S} \text{ and } R) = 0.5$;
$P(\overline{S} \text{ and } R) = 0.35$

$P(\overline{S}|R) = P(\overline{S} \text{ and } R)/P(R) = 0.35/0.50 = \underline{0.7}$

f. <u>No</u>. Independent events can intersect, therefore R and S are not mutually exclusive events.

4.85 a. P(both red) = (5/8)·(4/7) = <u>20/56</u>

b. P(one of each color) = (5/8)·(3/7) + (3/8)·(5/7)
= <u>30/56</u>

c. P(both white) = (3/8)·(2/7) = <u>6/56</u>

4.87 a.

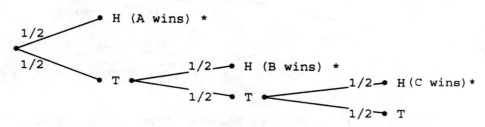

P(A wins on 1st turn) = <u>1/2</u>

P(B wins on 1st turn) = P(A does not)·P(B wins)
= (1/2)·(1/2) = <u>1/4</u>

P(C wins on 1st turn) = P(A does not)·P(B does not)·P(C wins)
= (1/2)·(1/2)·(1/2) = <u>1/8</u>

b.　　　　　　　　　1st try　　　　　　　　　2nd try

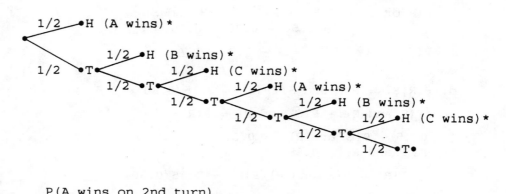

P(A wins on 2nd turn)
　　= P(A not on 1st)·P(B not)·P(C not)·P(A wins on 2nd)
　　= (1/2)·(1/2)·(1/2)·(1/2) = 1/16

P(A wins on 1st try or 2nd try) = 1/2 + 1/16 = <u>9/16</u>

P(B wins on 1st try or 2nd try) = 1/4 + 1/32 = <u>9/32</u>

P(C wins on 1st try or 2nd try) = 1/8 + 1/64 = <u>9/64</u>

4.89

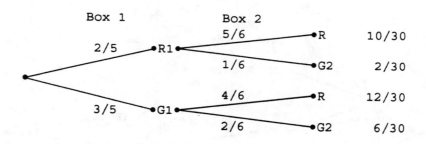

Let G2 = green ball is selected from Box 2 and R1 = red ball
is selected from box 1.

P(G2) = P[(R1 and G2) or (G1 and G2)]
　　= P(R1)·P(G2|R1) + P(G1)·P(G2|G1)
　　= (2/5)·(1/6) + (3/5)·(2/6) = 2/30 + 6/30 = <u>8/30</u>

4.91 a. $P(C) = 0.1$; $P(\overline{C}) = 1.0 - 0.1 = \underline{0.9}$

b. $P(D|C) = 1/4 = 0.25$; $P(\overline{D}|C) = 1 - P(D|C) = 1 - 1/4$
$$= 3/4 = 0.75;$$

$$P(C \text{ and } \overline{D}) = P(C) \cdot P(\overline{D}|C)$$
$$= (0.10) \cdot (0.75)$$
$$= \underline{0.075}$$

c. $P(C \text{ and } D) = P(C) \cdot P(D|C) = (0.1) \cdot (0.25) = \underline{0.025}$

CHAPTER EXERCISES

4.93 a. True. Law of Large Numbers.

b. True. Law of Large Numbers.

c. False. The irregularity will be the order of occurrence or the randomness of each individual outcome, not the overall summary of all 100 million outcomes. Further, the Law of Large Numbers says to expect about one-half of the 100 million to be H's and the other half to be T's.

d. False. The probability of a head occurring on any individual toss is the same as on any other toss, provided the tossing is done in a fair way.

4.95 a. P(blue eyes) = 90/300 = $\underline{0.30}$

b. P(yes) = 120/300 = $\underline{0.40}$

c. If independent; then $P(A \text{ and } B) = P(A) \cdot P(B)$

$P(A \text{ and } B) = 70/300 = 0.233$, and

$P(A) \cdot P(B) = (90/300) \cdot (120/300) = 0.12$; <u>not independent</u>

d. Blue eyes and brown eyes are mutually exclusive events. They are not complementary since not everyone was classified as having brown or blue eyes. Since they are mutually exclusive, they cannot be independent events.

4.97 a. <u>False.</u> If mutually exclusive, P(R or S) is found by adding 0.2 and 0.5.

   b. <u>True.</u> $0.2 + 0.5 - (0.2 \cdot 0.5) = 0.6$

   c. <u>False.</u> If mutually exclusive, P(R and S) must be equal to zero; there is no intersection.

   d. <u>False.</u> $0.2 + 0.5 = 0.7$, not 0.6

4.99 a. S = {GGG, GGR, GRG, GRR, RGG, RGR, RRG, RRR}

   b. P(exactly one R) = <u>3/8</u>

   c. P(at least one R) = <u>7/8</u>

> Rearrange probability formulas in order needed. Remember P(A and B) = P(A) · P(B|A) is the same as P(B and A) = P(B)· P(A|B) since A and B is the same as B and A.

4.101 a. P(A and B) = P(B)·P(A|B) = $(0.36) \cdot (0.88)$ = <u>0.3168</u>

   b. P(B|A) = P(A and B)/P(A) = $0.3168/0.68$ = <u>0.4659</u>

   c. <u>No.</u> P(A) does not equal P(A|B)

   d. <u>No.</u> P(A and B) does not equal 0.0

   e. It would mean that the two events "candidate wants job" and "RJB wants candidate" could not both happen.

4.103 Let A_i represent a 6 on the ith roll

   P(6 occurs first on the 5th roll) =

$$= P(\overline{A}_1 \text{ and } \overline{A}_2 \text{ and } \overline{A}_3 \text{ and } \overline{A}_4 \text{ and } A_5)$$
$$= P(\overline{A}_1) \cdot P(\overline{A}_2) \cdot P(\overline{A}_3) \cdot P(\overline{A}_4) \cdot P(A_5)$$
$$= (5/6) \cdot (5/6) \cdot (5/6) \cdot (5/6) \cdot (1/6) = \underline{0.080}$$

4.105 Let A = the individual will be between 20 and 29,
B = the individual will survive 3 years.

$P(A) = 0.31$ and $P(B|A) = 0.91$

$P(A \text{ and } B) = P(A) \cdot P(B|A) = 0.31 \cdot 0.91 = \underline{0.28}$

4.107 a. P(both damage free) = $(10/15) \cdot (9/14) = \underline{0.429}$

b. P(exactly one) = $(10/15) \cdot (5/14) + (5/15) \cdot (10/14)$
$= \underline{0.476}$

c. P(at least one)= $0.429 + 0.476 = \underline{0.905}$

4.109 P[(med or sh) and (mod or sev)] =
$= (90 + 121 + 35 + 54)/1000 = \underline{0.300}$

Note the wording: pink seedless denotes pink <u>and</u> seedless.   Use
formulas accordingly.

4.111 a. P(seedless) = $(10+20)/100 = \underline{0.30}$

b. P(white) = $(20+40)/100 = \underline{0.60}$

c. P(pink and seedless) = $10/100 = \underline{0.10}$

d. P(pink or seedless) = $(10+20+30)/100 = \underline{0.60}$

e. P(pink|seedless) = $0.10/0.30 = \underline{0.333}$

f. P(seedless|pink) = $0.10/0.40 = \underline{0.25}$

Exercise 4.113 involves a series of steps. This is a clue to use a tree diagram. Assign probabilities to the branches.

4.113 a.

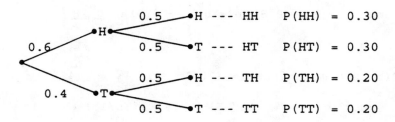

Coin A	Coin B	probability

0.5 ●H --- HH   P(HH) = 0.30

0.6 ●H

0.5 ●T --- HT   P(HT) = 0.30

0.5 ●H --- TH   P(TH) = 0.20

0.4 ●T

0.5 ●T --- TT   P(TT) = 0.20

b. P(both heads) = P(HH) = <u>0.30</u>

c. P(exactly one head) = P(HT or TH) = P(HT) + P(TH)
                         = 0.30 + 0.20 = <u>0.50</u>

d. P(neither is a head) = P(TT) = <u>0.20</u>

e. P(HH|A is a H) = <u>0.50</u>

f. P(HH|B is a head) = 0.30/(0.30+0.20) = <u>0.60</u>

g. P(H on A|exactly one H) = 0.30/(0.30+0.20) = <u>0.60</u>

4.115 Identify the proofreaders as A and B

a. P(error is found|each reads one page) =

= P[(A finds error on page read) or (B finds it)]

= P[(error is on page A reads and A finds it)
        or (error on page B reads and B finds it)]

= (0.5)(0.8) + (0.5)(0.8) = <u>0.80</u>

b. P(error is found|each read both pages)

= P(A finds error or B finds error)

= 0.8 + 0.8 - (0.8)(0.8) = <u>0.96</u>

c. P(error is found|both read one randomly selected page)

      = P[(error is on pg A reads and A finds it) or
               error is on pg B reads and B finds it)]

      = [(.5)(.8)] + [(.5)(.8)] - {[(.5)(.8)]·[(.5)(.8)]}

      = <u>0.64</u>

PROBABILITY

BASIC PROPERTIES: $0 \leq$ each probability ≤ 1 (1)

$$\sum_{overS} P(A) = 1 \qquad (2)$$

Finding probabilities	From an equally likely sample space	By formula, given certain probabilities
P(A), any event A	$P(A) = \dfrac{n(A)}{n(S)}$ (3)	-does not apply-
P($\overline{A}$), complementary event	$P(\overline{A}) = \dfrac{n(\overline{A})}{n(S)}$ (4)	$P(\overline{A}) = 1.0 - P(A)$ (11)
Any 2 events, no special conditions or relations known: P(A|B), conditional event P(A or B), union of 2 events P(A and B), intersection of 2 events	$P(A\|B) = \dfrac{n(A \text{ and } B)}{n(B)}$ (5) $P(A \text{ or } B) = \dfrac{n(A \text{ or } B)}{n(S)}$ (6) $P(A \text{ and } B) = \dfrac{n(A \text{ and } B)}{n(S)}$ (7)	$P(A\|B) = \dfrac{P(A \text{ and } B)}{P(B)}$ (12) $P(A \text{ or } B) = P(A) + P(B) - P(A \text{ and } B)$ (13) $P(A \text{ and } B) = P(A) \cdot P(B\|A)$ (14)
2 events, known to be mutually exclusive P(A or B) P(A and B) P(A|B)	$P(A \text{ or } B) = \dfrac{n(A) + n(B)}{n(S)}$ (8)	$P(A \text{ or } B) = P(A) + P(B)$ (15) $P(A \text{ and } B) = 0$ (16) $P(A\|B) = 0$ (17)
2 events, known to be independent P(A and B) P(A|B)	$P(A \text{ and } B) = \dfrac{n(A \text{ and } B)}{n(S)}$ (9) $P(A\|B) = \dfrac{n(A \text{ and } B)}{n(B)} = \dfrac{n(A)}{n(S)}$ (10)	$P(A \text{ and } B) = P(A) \cdot P(B)$ (18) $P(A\|B) = P(A)$ (19)

Resulting Properties:

(20) If $P(A) + P(B) = P(A \text{ or } B)$; then A and B are mutually exclusive.

(21) If $P(A) \cdot P(B) = P(A \text{ and } B)$; then A and B are independent.

(22) If $P(A|B) = P(A)$, then A and B are independent.

(23) If $P(A \text{ and } B) = 0$, then A and B are mutually exclusive.

(24) If $P(A \text{ and } B) \neq 0$, then A and B are not mutually exclusive.

The Relationship between Independence and Mutually Exclusive

(25) If events are independent, then they are NOT mutually exclusive.

(26) If events are mutually exclusive, then they are NOT independent.

CHAPTER 5 ∇ PROBABILITY DISTRIBUTIONS (DISCRETE VARIABLES)

Chapter Preview

Chapter 5 combines the "ideas" of a frequency distribution from Chapter 2 with probability from Chapter 4. This combination results in a discrete probability distribution. The main elements of this type distribution will be covered in this chapter. The elements include:

1. discrete random variables
2. discrete probability distributions
3. the mean and standard deviation of a discrete probability distribution
4. binomial probability distribution
5. the mean and standard deviation of a binomial distribution.

SECTION 5.1 MARGIN EXERCISES

5.1   The random variable is the number of courses per student. The possible values for the random variable are x = 1, 2, 3, ..., n.

5.2   The random variable is the combined weight for books and supplies being carried to class. The random variable will be a numerical value between 0 and 30 pounds for most students.

5.3   The random variable in exercise 5.1 is discrete. It can assume a countable number of values, but cannot be 1.73. The random variable in exercise 5.2 is continuous. Weight is a measurement and can assume any value along a line interval including all possible fractions.

5.4 a. The variable "score" is discrete because it is a count of points. There are no fractional points.
   b. The variable "number of minutes to commute to work" is continuous because time is a measurement with fractional values.

SECTION 5.1 EXERCISES

Random Variable - a numerical quantity whose value depends on the conditions and probabilities associated with an experiment.

Discrete Random Variable - a numerical quantity taking on or having a finite or countably infinite number of values.

Use x to denote a discrete random variable. (x is often a count of something; ex. the number of home runs in a baseball game)

5.5   The random variable is the *number of children per family*.

The possible values for the random variable are x = 0, 1, 2, 3, ... , n; where n is the maximum number of children for any family in the community.   The variable is discrete.

5.7   The random variable is the *distance from center to arrow*.

The possible values for the random variable are x = 0 to n, where n = radius of the target, measured in inches, including all possible fractions.   The variable is continuous.

SECTION 5.2 MARGIN EXERCISES

5.9

x	0	1
P(x)	1/2	1/2

5.10

x	1	2	3	4	5	6
P(x)	1/6	1/6	1/6	1/6	1/6	1/6

5.11 a. 1/6 = 0.1667 or 16.7%

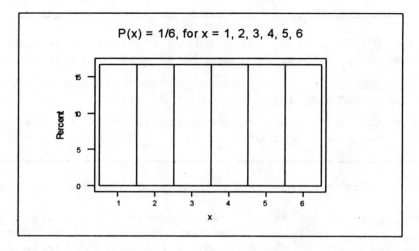

b. The distribution in (a) is uniform or rectangular.

5.12 a.

x	P(x)
0	0.05
1	0.12
2	0.15
3	0.25
4	0.21
5	0.10
6	0.05
7	0.04
8	0.02
9	0.01

b. 3 weeks = 21 days

$1/21 \approx 0.05$ = 5% shown on the
histogram

SECTION 5.2 EXERCISES

Probability distributions look very much like frequency
distributions. The probability P(x) takes the place of frequency.
The frequency f column contains integers (counts), whereas the
probability $P(x)$ column contains fractions or decimals between 0
and 1. The probability P(x) relates to relative frequency, the
frequency divided by the size of the data set.

...

The two main properties of a probability distribution are:

    1. $0 \leq$ each P(x) ≤ 1, each probability is a number between 0 and 1 inclusive.

    2. $\sum$ P(x) = 1, the sum of all the probabilities should be equal to 1.

Remember to always: 1. make sure each entry in the P(x) column is between 0 and 1, and

    2. sum your P(x) column and check that it is equal to 1.

Both properties <u>must</u> exist.

5.13

x	0	1	2	3
P(x)	0.20	0.30	0.40	0.10

Notice that each P(x) is a value between 0.0 and 1.0, and the sum of all P(x) values is exactly 1.0.

<u>Function Notation</u>

    P(x) $\Rightarrow$ an equation with x as its variable, which assigns probabilities to the corresponding or given values.

    P(0) $\Rightarrow$ replace x on the right side of the equation with 0 and evaluate.

    P(3) $\Rightarrow$ replace x on the right side of the equation with 3 and evaluate.

ex.: $P(x) = \dfrac{x + 1}{26}$ $P(0) = \dfrac{0 + 1}{26} = \dfrac{1}{26}$ $P(3) = \dfrac{3 + 1}{26} = \dfrac{4}{26}$

NOTE: Only evaluate P(x) for the x values in its domain, otherwise P(x) = 0. The domain of a variable is the specified set of replacements (x-values).

5.15 a.

x	P(x)
1	0.12
2	0.18
3	0.28
4	0.42
Σ	1.00

P(x) is a probability function:

1. Each P(x) is a value between 0 and 1.

2. The sum of the P(x)'s is 1.

b.

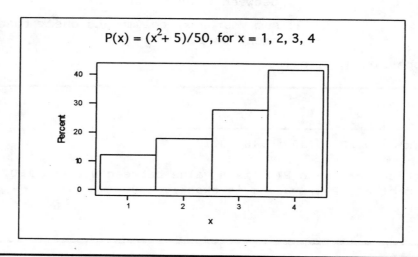

$$P(x) = (x^2 + 5)/50, \text{ for } x = 1, 2, 3, 4$$

* What shape distribution does the histogram in exercise 5.15b depict? (See the bottom of the next page for answer.)

5.17 No, the percentages are the percentage of each age group; that is, 1.7% of 12 year olds smoke; therefore 98.7% of the 12 year olds do not smoke. Age is an identifier for the categories, not used as a random variable.

For more information on the MINITAB command to generate discrete data according to a probability distribution, see JES2-p216. Note: The higher the probability, the more often the number will be generated.

5.19 a. Everyone's generated values will be different. Listed here is one such sample.

```
2  2  3  2  2  5  3  3  2  2  2  1  4  3  3  1
5  1  3  3  3  4  3  3  5
```

b. Sample obtained
    x   rel.freq.

    1   0.12
    2   0.28
    3   0.40
    4   0.08
    5   0.12
  ALL   1.00

c. Given Distribution

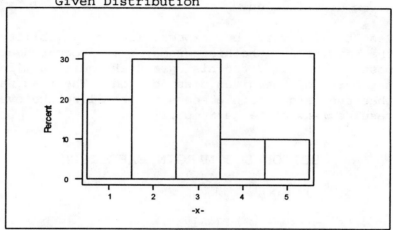

Sample Results

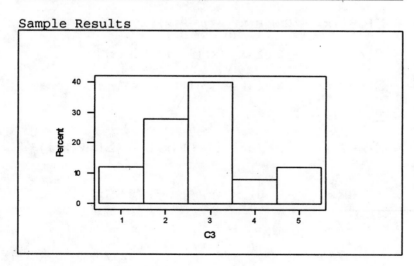

*(5.15b) A J-shaped distribution.

d. The distribution of the sample is somewhat similar to that of the given distribution. The two highest probabilities in the random data occurred at x = 2 and 3, matching the two highest probabilities for the given distribution.

e. Results will vary, but expect: occasionally a sample will have no 4 or no 5 in it, 2 or 3 is almost always the most frequent number, 4 or 5 is almost always the least frequent number, the histograms seem to vary but yet almost always look somewhat like the histogram of the given distribution.

f. Results will vary, but expect: little variability among the samples and the histograms, 4 or 5 occur nearly 10% most of the time, the histograms are quite similar to the histogram of the given distribution. The results indicate that the larger sample seems to stabilize the overall results (Law of Large Numbers).

SECTION 5.3 MARGIN EXERCISES

5.21 $\sigma^2 = \sum[(x - \mu)^2 \cdot P(x)]$

$= \sum[(x^2 - 2x\mu + \mu^2) \cdot P(x)]$

$= \sum[x^2 \cdot P(x) - 2x\mu \cdot P(x) + \mu^2 \cdot P(x)]$

$= \sum[x^2 \cdot P(x)] - 2\mu \cdot \sum[x \cdot P(x)] + \mu^2 \cdot [\sum P(x)]$

$= \sum[x^2 \cdot P(x)] - 2\mu \cdot [\mu] + \mu^2 \cdot [1]$

$= \sum[x^2 \cdot P(x)] - 2\mu^2 + \mu^2$

$= \sum[x^2 \cdot P(x)] - \mu^2$ or $\sum[x^2 \cdot P(x)] - \{\sum[x \cdot P(x)]\}^2$

5.22 a.

x	P(x)	b) xP(x)	$x^2 P(x)$
1	1/6	1/6	1/6
2	2/6	4/6	8/6
3	3/6	9/6	27/6

$\sum$	6/6 = 1.0	c) 14/6 = 2.33	36/6 = 6.0
	ck		

5.23 $\mu = \Sigma[xP(x)] = \underline{2.33}$

5.24 $\sigma^2 = \Sigma[x^2P(x)] - \{\Sigma[xP(x)]\}^2 = 6.0 - \{2.3333\}^2 = \underline{0.55556}$

5.25 $\sigma = \sqrt{\sigma^2} = \sqrt{0.55556} = \underline{0.745}$

5.26 The sum of the number values, once each. Nothing of any meaning.

SECTION 5.3 EXERCISES

The mean and standard deviation of a probability distribution are μ and σ respectively. They are parameters since we are using theoretical probabilities.

$$\mu = \Sigma[xP(x)] \quad \sigma^2 = \Sigma[x^2P(x)] - \{\Sigma[xP(x)]\}^2 \Rightarrow \sigma = \sqrt{\sigma^2}$$
$$\text{OR}$$
$$\sigma^2 = \Sigma[x^2P(x)] - \mu^2 \qquad \Rightarrow \sigma = \sqrt{\sigma^2} \text{ (easier formula)}$$

5.27

x	P(x)	xP(x)	x²P(x)
1	4/10	4/10	4/10
2	3/10	6/10	12/10
3	2/10	6/10	18/10
4	1/10	4/10	16/10
Σ	10/10 = 1.0 ck	20/10 = 2.0	50/10 = 5.0

$\mu = \Sigma[xP(x)] = \underline{2.0}$

$\sigma^2 = \Sigma[x^2P(x)] - \{\Sigma[xP(x)]\}^2 = 5.0 - \{2.0\}^2 = 1.0$

$\sigma = \sqrt{\sigma^2} = \sqrt{1.0} = \underline{1.0}$

5.29 a.

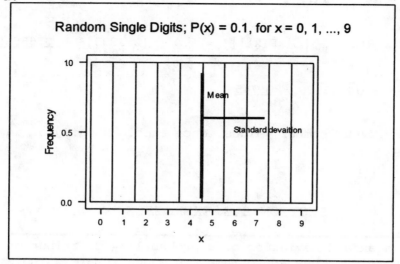

Random Single Digits; P(x) = 0.1, for x = 0, 1, ..., 9

* What shape distribution does the histogram in exercise 5.29a depict? Is the mean where you would expect it? (See bottom of the next page for answer.

b.

x	P(x)	xP(x)	x²P(x)
0	0.1	0.0	0.0
1	0.1	0.1	0.1
2	0.1	0.2	0.4
3	0.1	0.3	0.9
4	0.1	0.4	1.6
5	0.1	0.5	2.5
6	0.1	0.6	3.6
7	0.1	0.7	4.9
8	0.1	0.8	6.4
9	0.1	0.9	8.1
Σ	1.0 ck	4.5	28.5

$\mu = \Sigma[xP(x)] = \underline{4.5}$

$\sigma^2 = \Sigma[x^2P(x)] - \{\Sigma[xP(x)]\}^2 = 28.5 - \{4.5\}^2 = 8.25$

$\sigma = \sqrt{\sigma^2} = \sqrt{8.25} = \underline{2.87}$

c. See graph in part (a).

-- 112 --

d. $\mu \pm 2\sigma = 4.5 \pm 2(2.87) = 4.5 \pm 5.74$ or -1.24 to 10.24

The interval from -1.24 to 10.24 contains all the x-values of this probability distribution; <u>100%</u>

** How does the 100% from exercise 5.29d compare to Chebyshev's Theorem? (See bottom of this page for answer.)

5.31 a.

x	P(x)	xP(x)	x²P(x)
1	0.6	0.6	0.6
2	0.1	0.2	0.4
3	0.1	0.3	0.9
4	0.1	0.4	1.6
5	0.1	0.5	2.5
Σ	1.0 ck	2.0	6.0

$\mu = \Sigma[xP(x)] = \underline{2.0}$

$\sigma^2 = \Sigma[x^2P(x)] - \{\Sigma[xP(x)]\}^2 = 6.0 - \{2.0\}^2 = 2.0$

$\sigma = \sqrt{\sigma^2} = \sqrt{2.0} = \underline{1.4}$

b. $\mu - 2\sigma = 2.0 - 2(1.4) = -0.8$

$\mu + 2\sigma = 2.0 + 2(1.4) = 4.8$

The interval from -0.8 to 4.8 encompasses the numbers <u>1, 2, 3 and 4.</u>

c. The total probability associated with these values of x is <u>0.9</u>.

---

* How does this value of 0.9 in exercise 5.31 compare with Chebyshev's theorem? (See the bottom of the next page for answer.)

---

*(5.29a) A uniform distribution. The mean is exactly in the center since all outcomes are equally likely.
**(5.29d) Chebyshev's theorem states that for any shape distribution, at least 75% of the data is within 2 standard deviations of the mean. 100% is well over the minimum limit of 75%.

-- 113 --

5.33 Each question is in itself a separate trial with its own outcome having no effect on the outcomes of the other questions.

5.34 There are four different ways that one correct and three wrong answers can be obtained in four questions, each with the same probability. The sum of the 4 probabilities is the same value as 4 times one of them.

5.35 The 1/3 is the probability of success for each question, i.e., the probability of choosing the right answer from the 3 choices. The 4 is the number of independent trials, i.e., the number of questions.
The expected average would be the sample size times the probability of success; if the probability of guessing one answer correctly is 1/3, then it seems reasonable that on the average one should be able to guess 1/3 of all questions correctly.

5.36 Answers will vary but will be in the form of a frequency distribution, where x is the number of correct answers.

x	f
0	
1	
2	
3	
4	
Σ	20

5.37 Property 1: One trial is the flip of <u>one coin</u>, repeated <u>n = 50</u> times. The trials are independent because the probability of a head on any one toss has no effect on the probabilities for the other tosses.

---

*(5.31) Chebyshev's theorem states that for any shape distribution, at least 75% of the data is within 2 standard deviations of the mean. 90% is well over the minimum limit of 75%.

Property 2: Two outcomes on each trial:
                  success = H, heads
                  failure = T, tails
Property 3: p = P(heads) = 1/2 and
                  q = P(tails) = 1/2 [p+q=1]
Property 4: x = the number of heads for the experiment and
                  can be any integer value from 0 to 4.

5.38 a. $4! = 4 \cdot 3 \cdot 2 \cdot 1 = \underline{24}$

b. $\binom{4}{3} = \dfrac{4!}{3! \ 1!} = \dfrac{4 \cdot 3 \cdot 2 \cdot 1}{3 \cdot 2 \cdot 1 \cdot 1} = 4$

5.39 $P(x) = \binom{3}{x}(0.5)^x (0.5)^{3-x}$

$P(0) = \binom{3}{0}(0.5)^0 (0.5)^3 = 1(1)(0.125) = 0.125$

$P(2) = \binom{3}{2}(0.5)^2 (0.5)^1 = 3(0.25)(0.5) = 0.375$

$P(3) = \binom{3}{3}(0.5)^3 (0.5)^0 = 1(0.125)(1) = 0.125$

5.40 a. $P(4) = \binom{5}{4}\left(\dfrac{1}{4}\right)^4 \left(\dfrac{3}{4}\right)^1 = 5(0.0039)(0.75) = 0.0146$

$P(5) = \binom{5}{5}\left(\dfrac{1}{4}\right)^5 \left(\dfrac{3}{4}\right)^0 = 1(0.0009766)(1) = 0.00098$

b.

x	P(x)
0	0.2373
1	0.3955
2	0.2637
3	0.0879
4	0.0146
5	0.00098
Σ	0.99998

$\Sigma P(x) = 0.99998 \approx 1$ (round-off error)

$0 \leq$ each $P(x) \leq 1$

5.41 P(replacement) = risk = 1 - (0.886 + 0.107) = <u>0.007</u> = <u>0.7%</u>

5.42 P(x = 3|B(12, 0.30)) = <u>0.240</u>

SECTION 5.4 EXERCISES

FACTORIALS

$n! = n(n-1)(n-2)...1$ ex.: $3! = 3\cdot2\cdot1 = 6$

$0! = 1$ (this is defined this way so that the algebra of factorials will work)

$$\binom{n}{x} = \binom{n(trials)}{n(successes)\ n(failures)} = \frac{n!}{x!(n-x)!}$$

ex.: $\binom{8}{3} = \binom{8 trials}{3 successes\ 5 failures} = \frac{8!}{3!5!} = \frac{8\cdot7\cdot6\cdot5\cdot4\cdot3\cdot2\cdot1}{3\cdot2\cdot1\cdot5\cdot4\cdot3\cdot2\cdot1}$ or

$$= \frac{8\cdot7\cdot6\cdot5!}{3\cdot2\cdot1\cdot5!} = 56$$

EXPONENTS

$b^n = b\cdot b\cdot b\cdots$ (n times)   ex.: $.2^3 = (.2)(.2)(.2) = .008$

NOTE: See additional information about factorial notation in Review Lessons.

5.43 a. $4! = 4\cdot3\cdot2\cdot1 = \underline{24}$

b. $7! = 7\cdot6\cdot5\cdot4\cdot3\cdot2\cdot1 = \underline{5,040}$

c. $0! = \underline{1}$ (by definition)

d. $\dfrac{6!}{2!} \qquad \dfrac{6 \cdot 5 \cdot 4 \cdot 3 \cdot 2 \cdot 1}{2 \cdot 1} = 6 \cdot 5 \cdot 4 \cdot 3 = \underline{360}$

e. $\dfrac{5!}{3! \cdot 2!} = \dfrac{5 \cdot 4 \cdot 3 \cdot 2 \cdot 1}{3 \cdot 2 \cdot 1 \cdot 2 \cdot 1} = \underline{10}$

f. $\dfrac{6 \cdot 5 \cdot 4 \cdot 3 \cdot 2 \cdot 1}{4 \cdot 3 \cdot 2 \cdot 1 \cdot 2 \cdot 1} = \underline{15}$

g. $(0.3)^4 = (0.3)(0.3)(0.3)(0.3) = \underline{0.0081}$

h. $\dfrac{7 \cdot 6 \cdot 5 \cdot 4 \cdot 3 \cdot 2 \cdot 1}{3 \cdot 2 \cdot 1 \cdot 4 \cdot 3 \cdot 2 \cdot 1} = \underline{35}$

i. $\dfrac{5!}{2! \cdot 3!} = \dfrac{5 \cdot 4 \cdot 3 \cdot 2 \cdot 1}{2 \cdot 1 \cdot 3 \cdot 2 \cdot 1} = \underline{10}$

j. $\dfrac{3!}{0! \cdot 3!} = \dfrac{3 \cdot 2 \cdot 1}{1 \cdot 3 \cdot 2 \cdot 1} = \underline{1}$

$\dbinom{4}{1}(0.2)^1(0.8)^3 = \dbinom{4}{1} \cdot (0.2)^1 \cdot (0.8)^3$ The use of the multiplication dot is optional. They are sometimes used to emphasize that each of the three parts to a binomial must be evaluated separately first, then multiplication can take place.

---

k. $4 \cdot (0.2)(0.8)(0.8)(0.8) = \underline{0.4096}$

l. $1 \cdot 1 \cdot (0.7)^5 = \underline{0.16807}$

```
BINOMIAL EXPERIMENTS must have:

    1. n independent repeated trials
        a) n - the number of times the trial is repeated
        b) independent - the probabilities of the outcomes
            remain the same throughout the entire experiment

    2. two possible outcomes for each trial
        a) success - the outcome or group of outcomes that is
            the focus of the experiment
        b) failure - the outcome or group of outcomes not
            included in success

    3. p = probability of success on any one trial
       q = probability of failure on any one trial (q = 1 - p)

    4. x = number of successes when the experiment of all n
       trials is completed.  x can range in value from
       0 through n.  However, when the experiment is
       completed, x will have exactly one value, that is, the
       number of successes that occurred.
```

5.45 Binomial properties:

$n = 100$ trials (shirts),
two outcomes (first quality or irregular),
$p = P(\text{irregular})$,
$x = n(\text{irregular})$; any integer value from 0 to 100.

5.47 a. x is not a binomial random variable because the trials are
         not independent. The probability of success (get an ace)
         changes from trial to trial. On the first trial it is
         4/52. The probability of an ace on the second trial
         depends on the outcome of the first trial; it is 4/51 if
         an ace is not selected, and it is 3/51 if an ace was
         selected. The probability of an ace on any given trial
         continues to change when the experiment is completed
         without replacement.

b. x is a binomial random variable because the trials are
    independent. n = 4, the number of trials; two outcomes,
    success = ace and failure = not ace; p = P(ace) = 4/52 and
    q = P(not ace) = 48/52; x = n(aces drawn in 4 trials) and
    could be any number 0, 1, 2, 3 or 4. Further, the
    probability of success (get an ace) remains 4/52 for each
    trial throughout the experiment, as long as the card drawn
    on each trial is replaced before the next trial occurs.

5.49 a.

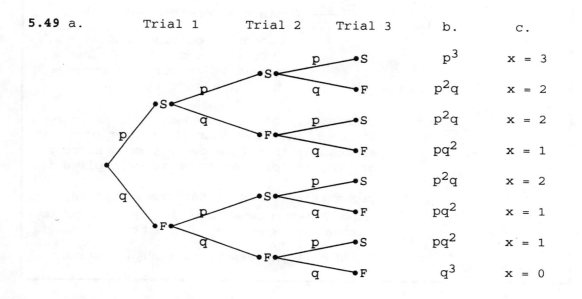

	Trial 1	Trial 2	Trial 3	b.	c.
				p^3	x = 3
				p^2q	x = 2
				p^2q	x = 2
				pq^2	x = 1
				p^2q	x = 2
				pq^2	x = 1
				pq^2	x = 1
				q^3	x = 0

e. $P(x) = \binom{3}{x} p^x q^{3-x}$, for x = 0, 1, 2, 3

BINOMIAL PROBABILITY FUNCTION

$$P(x) = \binom{n}{x} p^x q^{n-x} \text{ for } x = 0, 1, 2, \ldots n$$

where: $P(x)$ = probability of x successes

n   = the number of independent trials

$$\binom{n}{x} = \frac{n!}{x!(n-x)!} = \text{binomial coefficient} = \text{the}$$

number of combinations of
successes and failures that
result in exactly x successes
in n trials.

p^x = probability of x successes, that is, $p \cdot p \cdot p \cdots$,
x times. Remember x is the number of
successes, therefore every time a success
occurs, the probability p is multiplied in.

q^{n-x} = probability of (n-x) failures, that is,
$q \cdot q \cdot q \cdots$, (n-x) times. This is the probability
for "all of the rest of the trials."

Check: The sum of the exponents should equal n.

Exercise 5.51, parts a and c show more detailed solutions.   Review
factorials on ST-p116 of this manual, if necessary.

5.51   a. $\binom{4}{1} (0.3)^1 (0.7)^3 = \dfrac{4!}{1!\,3!} (0.3)^1 (0.7)^3$

$$= 4(0.3)(0.343) = \underline{0.4116}$$

b. $\binom{3}{2} (0.8)^2 (0.2)^1 = \underline{0.384}$

c. $\binom{2}{0} (1/4)^0 (3/4)^2 = \dfrac{2!}{0!\,2!} (1/4)^0 (3/4)^2 = 1(1)(9/16) = \underline{0.5625}$

d. $\binom{5}{2}(1/3)^2(2/3)^3 = \underline{0.329218}$

e. $\binom{4}{2}(0.5)^2(0.5)^2 = \underline{0.375}$

f. $\binom{3}{3}(1/6)^3(5/6)^0 = \underline{0.0046296}$

Use Table 2, the Binomial Probability Table, to find the needed
probabilities. Locate n and p, then the particular x or x's. If
more than one x is needed, add the probabilities.

5.53 By inspecting the function we see the binomial properties:

1. n = 5,

2. p = 1/2 and q = 1/2 (p + q = 1),

3. The two exponents x and 5-x add up to n = 5, and

4. x can take on any integer value from zero to n = 5;

   therefore it is binomial.

By inspecting the probability distribution:

x	T(x)	
0	1/32	It is a probability distribution.
1	5/32	
2	10/32	1. Each T(x) is between 0 and 1.
3	10/32	
4	5/32	2. $\sum T(x) = 1.0$
5	1/32	
$\sum$	32/32 = 1.0	

$$T(x) = \binom{5}{x}(1/2)^x(1/2)^{5-x} \text{ for } x = 0, 1, \ldots, 5$$

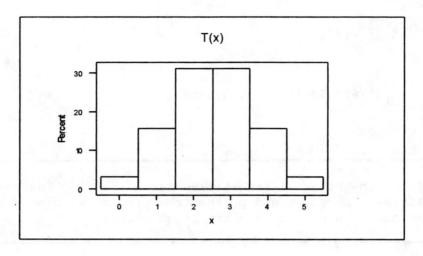

5.55 $P(x = 8, 9, 10 \mid B(n = 10, p = 0.90)) = P(8) + P(9) + P(10)$
                               $= 0.194 + 0.387 + 0.349 = \underline{0.930}$

5.57 $P(x = 8, 9, 10 \mid B(n = 11, p = 0.90)) = P(8) + P(9) + P(10)$
                               $= 0.071 + 0.213 + 0.384 = \underline{0.668}$

Terminology in probability problems
For a binomial problem with **n = 10**:

      at least 5 successes $\Rightarrow$ x = 5,6,7,8,9 or 10
      at most 5 successes $\Rightarrow$ x = 0,1,2,3,4 or 5

      at most 9 successes $\Rightarrow$ x = 0,1,2,3,4,5,6,7,8 or 9;
          Since **P(0) + P(1) + ... + P(9)** + P(10) = 1, then
              P(0) + P(1) + ... + P(9) = 1 - P(10).
          Therefore use: P(at most 9) = 1 - P(10).

      at least 2 successes $\Rightarrow$ x = 2,3,4,5,6,7,8,9,10
          Since P(0) + P(1) + **P(2) + P(3) +...+ P(9) + P(10)** = 1,
          then P(2) + P(3) +...+ P(9) + P(10) = 1 - [P(0) + P(1)].
          Therefore use: P(at least 2) = 1 - [P(0) + P(1)].

5.59   a. $P(x = 5|B(n = 5, p = 0.90)) = \underline{0.590}$

      b. $P(x = 4,5|B(n = 5, p = 0.90)) = 0.328 + 0.590 = \underline{0.918}$

5.61   P(shut down) = $P(x \geq 2)$, where x represents the number
     defective in the sample of n = 10.

     $P(x \geq 2) = 1.0 - [P(x = 0) + P(x = 1)]$

        $P(x = 0) = \binom{10}{0}(0.005)^0(0.995)^{10} = 0.9511$

        $P(x = 1) = \binom{10}{1}(0.005)^1(0.995)^9 = 0.0478$

     $P(x \geq 2) = 1.0 - [0.9511 + 0.0478] = \underline{0.0011}$

5.63   a. $P(x = 0|B(n = 15, p = P(\text{infect}) = 1/20)) = \underline{0.463}$

      b. $P(x \geq 1) = 1.0 - P(x = 0) = 1.000 - 0.463 = \underline{0.537}$

5.65 $P(x = 3|B(n = 6, p = 0.25)) = \binom{6}{3}(0.25)^3(0.75)^3 = \underline{0.132}$

5.67 a. $P(x = 3 | B(n = 5, p = 0.45)) = \binom{5}{3}(0.45)^3(0.55)^2 = \underline{0.2757}$

b. $P(x = 7 | B(n = 15, p = 0.45)) = [(0.177 + 0.196)/2] = \underline{0.1865}$

c. $P(x \geq 7 | B(n = 15, p = 0.45)) = [0.1865 + 0.157 + 0.107 + 0.058 + 0.0245 + 0.008 + 0.0015] = \underline{0.5425}$

d. $P(x \leq 7 | B(n = 15, p = 0.45)) = [0.0025 + 0.0125 + 0.0385 + 0.0845 + 0.139 + 0.18 + 0.1865] = \underline{0.6435}$

MINITAB commands found in JES2-p232 will save time in answering Exercise 5.69

5.69 MINITAB commands: CDF 2 C2;
BINOmial 50 0.023.

$P(x \leq 2 | B(n = 50, p = 0.023)) = \underline{0.8922}$

5.71 Commands needed: PDF C1 C2;
BINOmial 30 0.35.

x	P(x)	x	P(x)	x	P(x)
1*	0.0000	8	0.1009	15	0.0351
2	0.0003	9	0.1328	16	0.0177
3	0.0015	10	0.1502	17	0.0079
4	0.0056	11	0.1471	18	0.0031
5	0.0157	12	0.1254	19	0.0010
6	0.0353	13	0.0935	20	0.0003
7	0.0652	14	0.0611	21*	0.0001

* any other probabilities are each less than 0.00005

5.73 The number of defective items should be fairly small and therefore easier to count.

5.74 μ = np = 30·0.6 = <u>18</u>

$\sigma = \sqrt{npq} = \sqrt{30\cdot0.6\cdot0.4} = \sqrt{7.2} = 2.68 = \underline{2.7}$

5.75   a. μ = np =11·0.05 = <u>0.55</u>

$\sigma = \sqrt{npq} = \sqrt{11\cdot0.05\cdot0.95} = \sqrt{0.5225} = 0.7228 = \underline{0.72}$

b.

x	P(x)	xP(x)	x²P(x)
0	0.569	0	0
1	0.329	0.329	0.329
2	0.087	0.174	0.348
3	0.014	0.042	0.126
4	0.001	0.004	0.016
5	0+	0	0
Σ	1.0	0.549	0.819

The probabilities for x = 6 through 11 are all 0+.

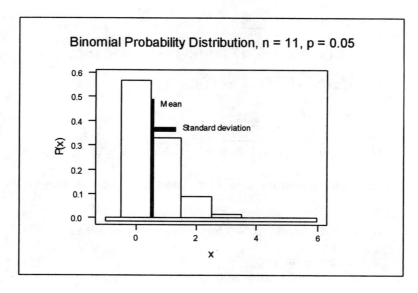

c. See histogram in (b).

For a <u>binomial distribution</u> the <u>mean</u> can be calculated using $\mu = np$ and the <u>standard deviation</u> by using $\sigma = \sqrt{npq}$.

Remember for <u>any type</u> discrete probability distribution:
$\mu = \Sigma[xP(x)]$ and $\sigma = \sqrt{\sigma^2}$ where $\sigma^2 = \Sigma[x^2P(x)] - [\Sigma[xP(x)]]^2$

Both formulas work for the binomial, however np and $\sqrt{npq}$ are quicker and less prone to computational error.

5.77 a.

x	T(x)	xT(x)	x^2T(x)
0	1/32	0/32	0/32
1	5/32	5/32	5/32
2	10/32	20/32	40/32
3	10/32	30/32	90/32
4	5/32	20/32	80/32
5	1/32	5/32	25/32
Σ	32/32 = 1.0	80/32 = 2.5	240/32 = 7.5

$\mu = \Sigma[xP(x)] = \underline{2.5}$

$\sigma^2 = \Sigma[x^2P(x)] - \{\Sigma[xP(x)]\}^2 = 7.5 - \{2.5\}^2 = 1.25$

$\sigma = \sqrt{\sigma^2} = \sqrt{1.25} = 1.118 = \underline{1.1}$

b. $\mu = np = 5 \cdot 0.5 = \underline{2.5}$

$\sigma = \sqrt{npq} = \sqrt{5 \cdot 0.5 \cdot 0.5} = \sqrt{1.25} = 1.118 = \underline{1.1}$

c. The same answers are obtained by both methods.

5.79 $\mu = np = 200$ and $\sigma = \sqrt{npq} = 10$, therefore:

$npq = 100$

$200q = 100$

$q = 100/200 = 0.5$

$p = 1 - q = \underline{0.5}$

$n = 200/0.5 = \underline{400}$

5.81 a. If assume x is binomial (approximately):

$$P(x) = \binom{5}{x} (0.75)^x (0.25)^{5-x} \text{ for } x = 0, 1, \ldots, 5$$

x	P(x)
0	0.00098
1	0.01465
2	0.08789
3	0.26367
4	0.39551
5	0.23730
Σ	1.00000

b. $P(x) = \binom{5}{x} (0.75)^x (0.25)^{5-x}$ for x = 0, 1, ... , 5

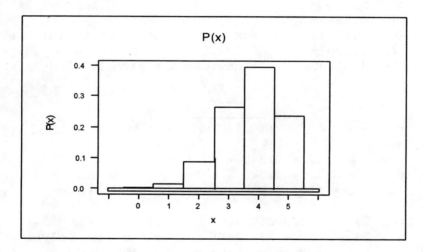

c. $\mu = np = 5 \cdot 0.75 = \underline{3.75}$

$\sigma = \sqrt{npq} = \sqrt{5 \cdot 0.75 \cdot 0.25} = \sqrt{0.9375} = 0.968 = \underline{0.97}$

5.83 a.

x	P(x)
0	0.886385
1	0.107441
2	0.005969
3	0.000201
4*	0.000005

*the rest are each less than 0.0000005

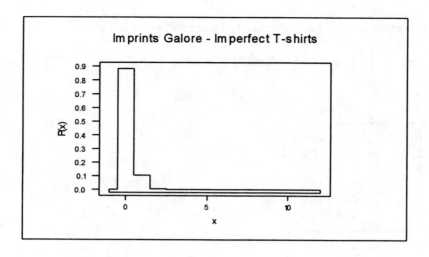

b. $P(x = 0 | B(n = 12, p = 0.01)) = \underline{0.886385}$

c. | x | cumP(x) |
    |---|---------|
    | 0 | 0.88638 |
    | 1 | 0.99383 |
    | 2 | 0.99979 |
    | 3* | 1.00000 |

$P(x = 0,1 | B(n = 12, p = 0.01)) = \underline{0.99383}$

*the rest are each 1.00000

d. $\mu = np = 12 \cdot (0.01) = \underline{0.12}$

$\sigma = \sqrt{npq} = \sqrt{12 \cdot (0.01) \cdot (0.99)} = 0.344674 = \underline{0.345}$

e. $\mu - \sigma = 0.12 - 0.345 = -0.225$ and
    $\mu + \sigma = 0.12 + 0.345 = 0.465$ includes only x = 0,
    therefore the proportion of the distribution is $\underline{0.88638}$

f. $\mu - 2\sigma = 0.12 - 2(0.345) = -0.57$ and
    $\mu + 2\sigma = 0.12 + 2(0.345) = 0.81$ includes only x = 0,
    therefore the proportion of the distribution is $\underline{0.88638}$

g. The two percentages do not agree with the empirical rule;
    the shape is J-shaped, not a normal shape. Part (f) does
    agree with Chebyshev's proportion of at least 0.75.
    Chebyshev's works for any shape distribution.

h.

C4	Count	Simulation	Expected
0	177	177/200 = 0.885	0.886385
1	21	21/200 = 0.105	0.107441
2	2	2/200 = 0.01	0.005969
3	0	0/200 = 0.00	0.000201
4	0	0/200 = 0.00	0.000005
N=	200		

	Simulation	Expected
Mean	0.125	0.12
St. Dev.	0.36059	0.345

i. All simulations were very close to the expected probabilities.

CHAPTER EXERCISES

5.85 a. A probability function is generally thought to be the *algebraic formula*, and a probability distribution is generally thought to be the *chart* listing the pairs of x and P(x) values. They are equivalent, just two very different-looking ways to express this one concept.

b. A probability distribution relates to a population, whereas a frequency distribution relates to a sample. To compare, a frequency distribution needs to become a relative frequency distribution by dividing each frequency by the sample size.

5.87 a.

x	f(x)
9	0.25
10	0.25
11	0.25
12	0.25
Σ	1.00

f(x) is a probability function since:

i) $0 \leq$ each f(x) ≤ 1
ii) $\Sigma f(x) = 1.0$

b.

x	f(x)
1	1.00
2	0.50
3	0.00
4	-0.50
Σ	1.00

f(x) is NOT a probability function since:

f(x=4) is not between 0 and 1

c.

x	f(x)
0	1/25
1	3/25
2	7/25
3	13/25
Σ	24/25

f(x) is NOT a probability function since:

$\Sigma f(x) = 24/25$, not 1.0

5.89 a. P(buyer watched once before buying) = 0.27

b. Can not be answered; the information is about buyers, not about viewers. Viewers only become part of the information when they make a purchase.

c. P(buyer watched 3 or more times before buying)
        = 0.18 + 0.09 + 0.15 = 0.42

d. No

e. Yes; $\Sigma P(x) = 1$ and $0 \leq$ each $P(x) \leq 1$

f.

x	P(x)	xP(x)	x²P(x)
1	0.27	0.27	0.27
2	0.31	0.62	1.24
3	0.18	0.54	1.62
4	0.09	0.36	1.44
5	0.15	0.75	3.75
Σ	1.0	2.54	8.32

$\mu = \Sigma[xP(x)] = \underline{2.54}$

$\sigma^2 = \Sigma[x^2P(x)] - \{\Sigma[xP(x)]\}^2 = 8.32 - \{2.54\}^2 = 1.8684$

$\sigma = \sqrt{\sigma^2} = \sqrt{1.8684} = 1.3669 = \underline{1.37}$

5.91 a. $P(x = 0|B(n = 5, p = 0.35)) = (0.65)^5 = \underline{0.116}$

b. $P(x = 4, 5|B(n = 5, p = 0.35)) = P(4) + P(5)$

$$P(x = 4|B(n = 5, p = 0.35)) = \binom{5}{4}(0.35)^4(0.65)^1 = 0.0488$$

$$P(x = 5|B(n = 5, p = 0.35)) = \binom{5}{5}(0.35)^5(0.65)^0 = 0.0053$$

$P(x = 4, 5) = 0.0488 + 0.0053 = \underline{0.0541}$

5.93 Two-engine plane:
$P(\text{successful flight}) = P(x = 1, 2|B(n = 2, p = 0.95))$
$$= 0.095 + 0.902 = \underline{0.997}$$

Four-engine plane:
$P(\text{successful flight}) = P(x = 2, 3, 4|B(n = 4, p = 0.95))$
$$= 0.014 + 0.171 + 0.815 = \underline{1.000}$$

The four-engine plane has a higher probability of a successful flight.

5.95 a. $P(x \geq 10|B(n = 15, p = 0.30)) = P(x = 10, 11, 12, 13, 14, 15)$
$P(x \geq 10) = 0.003 + 0.001 + 4(0+) = \underline{0.004}$

b. The chance of finding 10 or more in a sample of 15 when $p = 0.30$ is very low.

5.97 $\sigma^2 = \sum x^2 P(x) - \mu^2$ (Formula 5-3b)

$100 = \sum x^2 P(x) - 50^2$ or $\sum x^2 P(x) = \underline{2600}$

5.99 Tool Shop: x = profit

x	P(x)	x·P(x)
100,000	0.10	10,000.0
50,000	0.30	15,000.0
20,000	0.30	6,000.0
-80,000	0.30	-24,000.0
Σ	1.00ck	7,000.0

mean profit = $\mu = \sum [x \cdot P(x)]$ = 7,000.0

Book Store: x = profit

x	P(x)	x·P(x)
400,000	0.20	80,000.0
90,000	0.10	9,000.0
-20,000	0.40	-8,000.0
-250,000	0.30	-75,000.0
Σ	1.00ck	6,000.0

mean profit = $\mu = \sum [x \cdot P(x)]$ = 6,000.0

The Tool Shop has a slightly higher mean profit.

CHAPTER 6 ∇ NORMAL PROBABILITY DISTRIBUTIONS

Chapter Preview

Chapter 6 continues the presentation of probability distributions started in Chapter 5. In this chapter, the random variable is a continuous random variable (versus a discrete random variable in Chapter 5); therefore, the probability distribution is a continuous probability distribution. There are many types of continuous distributions, but this chapter will limit itself to the most common, namely, the normal distribution. The main elements of a normal probability distribution to be covered are:
1. how probabilities are found
2. how they are represented
3. how they are used.

SECTION 6.2 MARGIN EXERCISES

6.1   <u>0.4147</u>

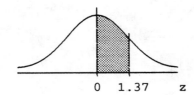

6.2

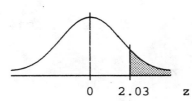

P(z > 2.03) =

0.5000 - 0.4788 = <u>0.0212</u>

6.3

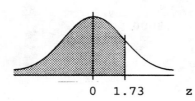

P(z < 1.73) =

0.5000 + 0.4582 = <u>0.9582</u>

6.4

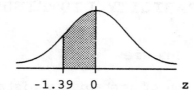

$P(-1.39 < z < 0.00) = \underline{0.4177}$

6.5

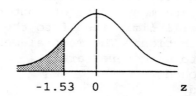

$P(z < -1.53) =$

$0.5000 - 0.4370 = \underline{0.0630}$

6.6

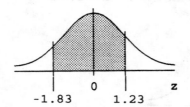

$P(-1.83 < z < 1.23) =$

$0.4664 + 0.3907 = \underline{0.8571}$

6.7

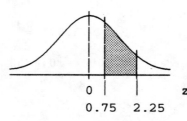

$P(0.75 < z < 2.25) =$

$0.4878 - 0.2734 = \underline{0.2144}$

6.8

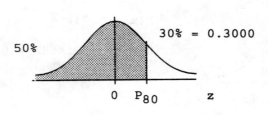

$z = \underline{0.84}$

$30\% = 0.3000$

$-- 134 --$

6.9

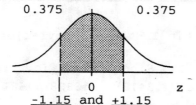

0.375   0.375 $0.75/2 = 0.375$

-1.15 and $+1.15$

SECTION 6.2 EXERCISES

Continuous Random Variable

      - a numerical quantity that can take on values in a certain interval

STANDARD NORMAL DISTRIBUTION

- bell shaped, symmetric curve
- $\mu = 0$, $\sigma = 1$
- distribution for the standard normal score z

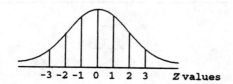

-3 -2 -1 0 1 2 3   **z values**

$z = 0 \Rightarrow \mu$

$z = 1 \Rightarrow \mu + 1\sigma$	$z = -1 \Rightarrow \mu - 1\sigma$
$z = 2 \Rightarrow \mu + 2\sigma$	$z = -2 \Rightarrow \mu - 2\sigma$
$z = 3 \Rightarrow \mu + 3\sigma$	$z = -3 \Rightarrow \mu - 3\sigma$

- area under the curve = 1

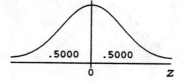

.5000   .5000

- symmetric, therefore

                                                          . . .

The Empirical Rule and Table 3

   ≈ 68% of the data lies within 1 standard deviation of the
      mean.
      - Note z = 1, gives .3413. Since symmetric, from z = -1
         to z = 1 would be 2(.3413) = .6828 ≈ 68%.

   ≈ 95% of the data lies within 2 standard deviations of the
      mean.
      - Note z = 2, gives .4772. Since symmetric, from z = -2
         to z = 2 would be 2(.4772) = .9544 ≈ 95%.

   ≈ 99.7% of the data lies within 3 standard deviations of
      the mean.
      - Note z = 3, gives .4987. Since symmetric, from z = -3
         to z = 3 would be 2(.4987) = .9974 ≈ 99.7%.

Draw a picture of a normal distribution and shade in the section
representing the area desired. Remember .5000 or 50% of the
distribution is on either side of μ or the z = 0.

6.11   a. <u>0.4032</u>

b. <u>0.3997</u>

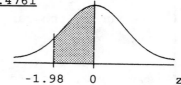

c. <u>0.4993</u>

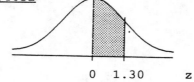

d. <u>0.4761</u>

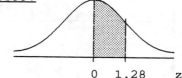

To find area in the right or left tail, subtract the Table 3 (Appendix B, JES2-p521) area (probability) for the given z value from 0.5000.

To find area that extends from one side of the mean (z = 0) through the tail of the other side, add the Table 3 area (probability) for the given z value to 0.5000.

6.13 a. <u>0.4394</u>   b. 0.5000 - 0.4394 = <u>0.0606</u>

c. 0.5000 + 0.4394 = <u>0.9394</u>   d. 2(0.4394) = <u>0.8788</u>

6.15 a. 0.3849 + 0.3888 = <u>0.7737</u>   b. 0.4599 + 0.4382 = <u>0.8981</u>

c. 0.4032 + 0.4951 = <u>0.8983</u>   d. 0.4998 - 0.1368 = <u>0.3630</u>

6.17 a. <u>0.5000</u>   b. 0.5000 - 0.3531 = <u>0.1469</u>

c. 0.5000 + 0.4893 = <u>0.9893</u>   d. 0.4452 + 0.5000 = <u>0.9452</u>

e. 0.5000 - 0.4452 = <u>0.0548</u>

6.19 a. <u>0.4906</u>   b. 0.4821 + 0.4904 = <u>0.9725</u>

c. 0.5000 - 0.0517 = <u>0.4483</u>   d. 0.5000 + 0.4306 = <u>0.9306</u>

Look at the *inside* of Table 3 (Appendix B, JES2-p521), and get as close as possible to the probability desired. Locate the position (row and column) on the outside of the table. This will be the corresponding z value. Remember the negative sign if the z value is to the left of µ.

NOTE: NORMAL CURVE   TABLE 3

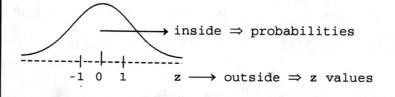

- inside ⇒ probabilities
- outside ⇒ z values

z ⟶ outside ⇒ z values

- inside ⇒
  probabilities
- outside ⇒
  z values

6.21   a. <u>1.14</u>   b. <u>0.47</u>   c. <u>1.66</u>

      d. <u>0.86</u>   e. <u>1.74</u>   f. <u>2.23</u>

Subtract the area desired from 0.5000 first. Table 3 (Appendix B, JES2-p521) works from 0 to a z value.
Remember that one half of the distribution is .5000.

6.23   a. <u>1.65</u>   b. <u>1.96</u>   c. <u>2.33</u>

Draw a picture of a normal distribution with the desired area shaded in.
Fill in the needed probabilities.
Locate the appropriate probability in Table 3 (Appendix B, JES2-p521).
Locate the corresponding z value.

6.25   40%   40%

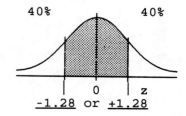

<u>-1.28</u> or <u>+1.28</u>

6.27   0.2500   0.2500   (The 50% is split in half.)

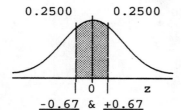

−0.67 & +0.67

6.29

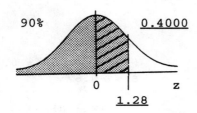

90%   0.4000   95%   0.4500

0   z   0   z

1.28   1.65

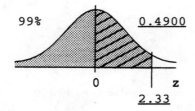

99%   0.4900

0   z

2.33

SECTION 6.3 MARGIN EXERCISES

6.31 $z = (x - \mu)/\sigma = (58 - 43/5.2) = \underline{2.88}$

6.32 $P(100 < x < 120) = P[(100 - 100)/16 < z < (120 - 100)/16]$
$= P[0.00 < z < 1.25] = \underline{0.3944}$

6.33 $P(x > 80) = P[z > (80 - 100)/16]$
$= P[z > -1.25]$
$= 0.3944 + 0.5000 = \underline{0.8944}$

6.34 15% = 0.1500 = 0.5000 - 0.3500
z closest to 0.3500 = 1.04
$z = (x - \mu)/\sigma$
1.04 = (x - 72)/13
13.52 = x - 72
x = 85.52 = <u>86</u>

6.35 a. z = (0.5505 - 0.5535)/0.0015 = -2.00

b. z = (0.5555 - 0.5535)/0.0015 = 1.33

6.36

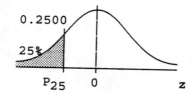

P(a < z < 0) = 0.2500
a = -0.67

$z = (x - \mu)/\sigma$
-0.67 = (x - 100)/16
-10.72 = x - 100
x = 89.28 = <u>89</u>

6.37

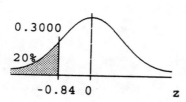

$z = (x - \mu)/\sigma$

-0.84 = (28,000 - μ)/1200

-1008 = 28,000 - μ

μ = <u>29,008</u>

6.38 Everyone's results will be different. Commands needed are:
        RANDom 100 C1;
        NORMal 50 12.
        PRINt C1

6.39 Everyone's results will be different. The corresponding
y-values will be in C2. Commands needed are:
        PDF C1 C2;
        NORMal 50 12.
        PRINt C1 C2

6.40 Everyone's results will be different. Each curve should be approximately normally distributed. Commands needed are:
        PLOT C2*C1;
        CONNect.

6.41 With 55 and 65 in C4 on MINITAB'S worksheet, enter the session commands: CDF C4 C5; NORM 50 12.
    $P(55 < x < 65) = 0.8944 - 0.6615 = 0.2329$

    $z = (55 - 50)/12 = 0.4167 = 0.42$
    $z = (65 - 50)/12 = 1.25$
    $P(0.42 < z < 1.25) = 0.3944 - 0.1628 = \underline{0.2316}$

    The difference is due to round-off error in the calculation of $z = 0.42$.

SECTION 6.3 EXERCISES

x is still used to denote a <u>continuous random variable</u>, but x is now referred to on an interval, not a single value. (ex.: a < x < b)
As before, any letter may be used; x is the most common.

<u>Applications of the Normal Distribution</u>
    1. Draw a sketch of the desired area, noting given μ and σ.

    2. Write the desired probability question in terms of the given variable - usually x (ex.: $P(x > 10)$).

    3. Transform the given variable x into a z value using
        $z = (x - \mu)/\sigma$.

    4. Rewrite the probability question using z
            (ex.: $P(z > *)$, where $*$ = calculated z value).

    5. Use Table 3 (Appendix B, JES2-p521) and find the probability.

Remember: Probabilities (areas) given in Table 3 are from $z = 0$
        to $z = \#$.

                                                              . . .

If the interval starts at z = 0 and continues either to the right or to the left, the probability is given in the table.

If the interval contains z = 0, add the probabilities.

If the interval extends from one positive z to another positive z, subtract the smaller probability from the larger probability (works the same with 2 negative z's).

If the interval does not contain z = 0 but does contain either the right or left tail, subtract the probability given in the table from 0.5000.

Reminder: When calculating z → use 2 decimal places.

 When calculating areas or probabilities → use 4 decimal places.

6.43   Use formula $z = (x - \mu)/\sigma$:

  a. $P[x > 250] = P[z > (250 - 210)/15] = P[z > 2.67]$
                $= 0.5000 - 0.4962 = 0.0038 = \underline{0.38\%}$

  b. $P[x < 150] = P[z < (150 - 210)/15] = P[z < -4.00]$
                $= 0.5000 - 0.49997 = 0.00003$
                $= \underline{0.003\%}$ (which is practically zero)

6.45 a. $P(40 < x < 60) = P[(40-52)/7 < z < (60-52)/7]$
                $= P[-1.71 < z < 1.14]$
                $= 0.4564 + 0.3729 = 0.8293 = \underline{82.93\%}$

  b. Perhaps $P(39.5 < x < 60.5)$ should be used in (a) since we do not use partial cents.
     If the distribution of x is not normal, the probabilities might not be very accurate.

6.47 a. $P[x < 2.0] = P[z < (2.0-3.7)/1.4]$
                $= P[z < -1.21]$
                $= 0.5000 - 0.3869 = \underline{0.1131}$

  b. $P[x > 6.0] = P[z > (6.0-3.7)/1.4]$
                $= P[z > 1.64]$
                $= 0.5000 - 0.4495 = \underline{0.0505}$

c.

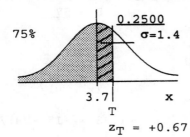

0.2500

75% σ=1.4

3.7   x
T

$z_T = +0.67$

$z = (x - \mu)/\sigma$

$0.67 = (T - 3.7)/1.4$

$T = (0.67)(1.4) + 3.7$

T = <u>4.64</u> <u>minutes</u>

6.49 a.

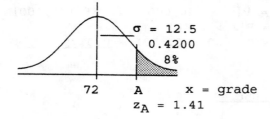

$\sigma = 12.5$
0.4200
8%

72   A   x = grade
$z_A = 1.41$

$z = (x - \mu)/\sigma$

$1.41 = (A - 72)/12.5$

$A = (1.41)(12.5) + 72$

A = <u>89.6</u>

b.

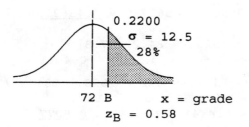

0.2200
$\sigma = 12.5$
28%

72 B   x = grade
$z_B = 0.58$

$z = (x - \mu)/\sigma$

$0.58 = (B - 72)/12.5$

$B = (0.58)(12.5) + 72$

B = <u>79.2</u>

c.

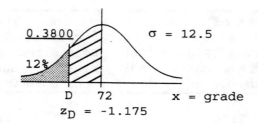

$$z = (x - \mu)/\sigma$$

$$-1.175 = (D - 72)/12.5$$

$$D = (-1.175)(12.5) + 72$$

$$D = \underline{57.3}$$

6.51

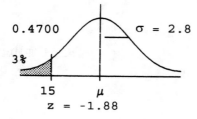

$$z = (x - \mu)/\sigma$$

$$-1.88 = (15 - \mu)/2.8$$

$$\mu = 15 - (-1.88)(2.8)$$

$$\mu = \underline{20.26}$$

6.53 a. $P(x < 15,000) = P[z < (15,000 - 26,362)/6,500]$
$$= P[z < -1.75]$$
$$= 0.5000 - 0.4599 = \underline{0.0401} = \underline{4\%}$$

b. $P(x > 40,000) = P[z > (40,000 - 26,362)/6,500]$
$$= P[z > 2.10]$$
$$= 0.5000 - 0.4821 = \underline{0.0179} = \underline{1.8\%}$$

6.55 a. $P(x < 525) = \underline{0.056241}$

b. $P(525 < x < 590) = 0.561785 - 0.056241 = \underline{0.505544}$

c. $P(x \geq 590) = 1.000000 - 0.561785 = \underline{0.438215}$

d. $P(x < 525) = P[z < (525 - 584.2)/37.3]$
$$= P[z < -1.59] = 0.5000 - 0.4441 = \underline{0.0559}$$

$P(525 < x < 590) = P[-1.59 < z < (590 - 584.2)/37.3]$
$$= P[-1.59 < z < 0.16]$$
$$= 0.4441 + 0.0636 = \underline{0.5077}$$

$P(x > 590) = P(z > 0.16) = 0.5000 - 0.0636 = \underline{0.4364}$

e. Round-off errors; specifically in (d) when z is calculated to the nearest hundredth and the rounded z score is used with Table 3.

6.57 Everyone's generated values will be different, but should have a mean and standard deviation close to 100 and 16, respectively, and be approximately normally distributed.

SECTION 6.4 MARGIN EXERCISES

6.59

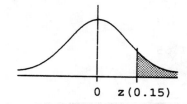

6.60

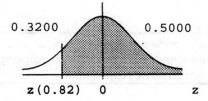

6.61 z(0.15) = <u>1.04</u>

6.62 z(0.82) = <u>-0.92</u>

SECTION 6.4 EXERCISES

Z - NOTATION

$z(\alpha) = z_\alpha$ = the z value that has α area to the right of it

1. Draw a picture.
2. Shade in desired α area, starting from the far right tail.
3. Based on the diagram and location, determine the z value using Table 3 (Appendix B, JES2-p521).

6.63 a. z(0.03)   b. z(0.14)   c. z(0.75)

d. z(0.13)   e. z(0.91)   f. z(0.82)

6.65 a. 1.96   b. 1.65   c. 2.33

Solve exercise 6.67 using Table 3 (Appendix B, JES2-p521).

6.67 a. 1.28, 1.65, 1.96, 2.05, 2.33, 2.58

Note the z values. They are the most common occurring z values. For that reason, Table 4, Part A, has been included in Appendix B (JES2-p522). Note Table 4, Part B for later use.

b. -2.58, -2.33, -2.05, -1.96, -1.65, -1.28

6.69 a. A is an area. z is 0.10 and the area to the right of z = 0.10 is 0.5000 - 0.0398 = <u>0.4602.</u>

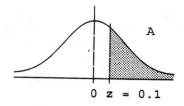

0 z = 0.1

b. B is a z-score. 0.10 is the area to the right of z = B. Use 0.4000 [0.5000 - 0.1000] to look up the z-score on Table 3 (Appendix B, JES2-p521). z = B = <u>1.28.</u>

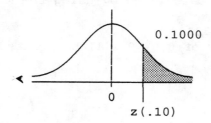

c. C is an area. z is -0.05 and the area to the right of z = -0.05 is 0.5000 + 0.0199 = <u>0.5199.</u>

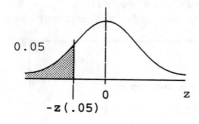

d. D is a z-score. D is to the left of zero [negative], use 0.4500 [0.5000 - 0.0500] to look it up; z = D = <u>-1.65.</u>

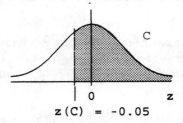

SECTION 6.5 MARGIN EXERCISES

6.71 np = (100)(0.02) = 2   nq = (100)(0.98) = 98
No, both np and nq must be greater than or equal to 5.

6.72 $P(x \leq 3 | B(25, 1/3)) \approx P(x < 3.5 | N(\mu, \sigma)) = P(z < -2.05) = \underline{0.202}$
$(\mu = np = 8.3333, \sigma = \sqrt{npq})$

$P(x \leq 3 | B(25, 1/3))$ can be calculated by using the MINITAB commands: CDF 3;
   BINOmial 25 0.33333.
The cumulative probability of $\underline{0.0149}$ will be printed on the session window.

SECTION 6.5 EXERCISES

Criteria for the Normal Approximation of the Binomial

1. np and n(1-p) must both be ≥ 5.

2. 0.5 must be added to and/or subtracted from the x values to allow for an interval. ex.: x = 5 $\Rightarrow$ 4.5 < x < 5.5
                                                    discrete   continuous

3. $\mu = np$ and $\sigma = \sqrt{npq}$

4. $z = (x - \mu)/\sigma$ and Table 3 (Appendix B, JES2-p521) are used to find probabilities.

6.73 a. Rule of thumb: np = 3 and nq = 7, therefore the approximation is not appropriate, since np < 5.
The distribution is slightly skewed, not symmetrical.

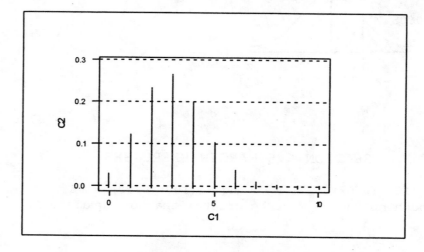

b. np = 0.5 and nq = 99.5, therefore the approximation is not appropriate, since np < 5. The distribution is J-shaped.

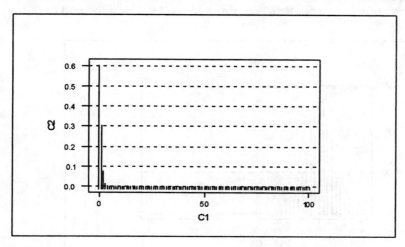

c. np = 50 and nq = 450, therefore the approximation is appropriate, since both np > 5 and nq > 5. The distribution appears to be approximately normal.

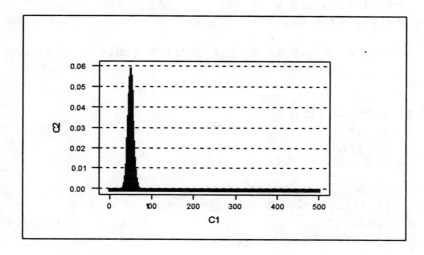

d. np = 10 and nq = 40, therefore the approximation is appropriate, since both np > 5 and nq > 5. The distribution appears to be approximately normal.

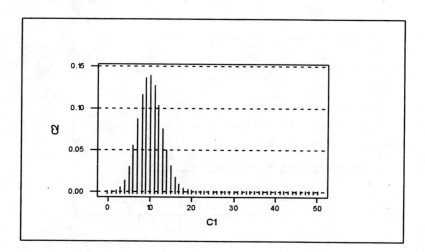

6.75 P(x = 6) = P(5.5 < x < 6.5)

= P[(5.5 - 7.2)/$\sqrt{2.88}$ < z < (6.5 - 7.2)/$\sqrt{2.88}$]

= P[-1.00 < z < -0.41]

= 0.3413 - 0.1591 = <u>0.1822</u>

P[x = 6|B(n=12, p = 0.6)] = <u>0.177</u> from Table 2 (Appendix B, JES2-p518)

6.77 P(x ≤ 8) = P(x < 8.5)

= P[z < (8.5 - 5.6)/$\sqrt{3.36}$]

= P[z < 1.58]

= 0.5000 + 0.4429 = <u>0.9429</u>

P[x ≤ 8|B(n = 14, p = 0.4)]

= P(0)+P(1)+P(2)+P(3)+P(4)+P(5)+P(6)+P(7)+P(8)

= 0.001+0.007+0.032+0.085+0.155+0.207+0.207+0.157+0.092

= <u>0.943</u> from Table 2 (Appendix B, JES2-p518)

6.79 Let x represent the number of patients in the 250 who will experience a side effect.

$\mu = np = (250)(0.05) = \underline{12.5}$
$\sigma = \sqrt{npq} = \sqrt{(250)(0.05)(0.95)} = \underline{3.45}$

$$P(x < 15.5) = P[z < (15.5 - 12.5)/3.45]$$
$$= P[z < 0.87]$$
$$= 0.5000 + 0.3078 = \underline{0.8078}$$

6.81 a. $\mu = np = (1024)(0.045) = 46.08 = \underline{46.1}$
$\sigma = \sqrt{npq} = \sqrt{(1024)(0.045)(0.955)} = \underline{6.63}$

$$P(35 \leq x \leq 50) = P(34.5 < x < 50.5)$$
$$= P[(34.5 - 46.1)/6.63 < z < (50.5 - 46.1)/6.63]$$
$$= P[-1.75 < z < 0.66]$$
$$= 0.4599 + 0.2454 = \underline{0.7053}$$

b. Commands needed: CDF C1 C3;
                     NORMal 46.1 6.63.
where C1 contains 34.5 and 50.5. The answer is found by subtraction, 0.7465 - 0.0401 = $\underline{0.7064}$

c. Commands needed: CDF C4 C5;
                     BINOmial 1024 0.045.
where C4 contains the numbers 34 and 50, since
$P(35 \leq x \leq 50) = P(x \leq 50) - P(x \leq 34)$. The answer is found by subtraction, 0.75145 - 0.03595 = $\underline{0.7155}$

CHAPTER EXERCISES

6.83 The range from z = -2 to z = +2 represents two standard deviations on either side of the mean. According to Chebyshev's theorem, there should be *at least 3/4* or *at least 0.75* of a distribution in this interval.

The area under a normal curve is 2(0.4772) or 0.9544.

6.85 a. $\underline{1.26}$ b. $\underline{2.16}$ c. $\underline{1.13}$

6.87   a. $P(|z| > 1.68) = P(z < -1.68) + P(z > +1.68)$
$\qquad\qquad\qquad = 2(0.5000 - 0.4535) = \underline{0.0930}$

   b. $P(|z| < 2.15) = P(-2.15 < z < +2.15)$
$\qquad\qquad\qquad = 2(0.4842) = \underline{0.9684}$

6.89   a. $\underline{1.175}$ or $\underline{1.18}$ b. $\underline{0.58}$

   c. $\underline{-1.04}$ d. $\underline{-2.33}$

6.91   a. $P[150 < x < 225] = P[(150-200)/25 < z < (225-200)/25]$
$\qquad\qquad\qquad\qquad = P[-2.00 < z < 1.00]$
$\qquad\qquad\qquad\qquad = 0.4772 + 0.3413 = \underline{0.8185}$

   b. $P[x > 250] = P[z > (250-200)/25]$
$\qquad\qquad\qquad = P[z > +2.00]$
$\qquad\qquad\qquad = 0.5000 - 0.4772 = \underline{0.0228}$

6.93

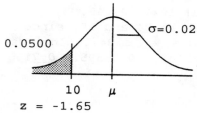

$z = (x - \mu)/\sigma$

$-1.65 = (10 - \mu)/0.02$

$\mu = 10 - (-1.65)(0.02)$

$\mu = \underline{10.033}$

6.95   a. The normal approximation is reasonable since both
$np = 7.5$ and $nq = 17.5$ are greater than 5.

   b. $\mu = np = (25)(0.3) = \underline{7.5}$

$\qquad \sigma = \sqrt{npq} = \sqrt{(25)(0.3)(0.7)} = \underline{2.29}$

6.97 a. Commands needed: SET C1
                                 0:50
                                 END
                                 PDF C1 C2;
                                 BINOmial 50 0.1.
                                 PRINt C1 C2

b. $P(x \leq 6)$ = 0.005154 + 0.028632 + 0.077943 + 0.138565
                 + 0.180904 + 0.184925 + 0.154104 = <u>0.77023</u>

c. $\mu = np = (50)(0.1) = \underline{5}$
    $\sigma = \sqrt{npq} = \sqrt{(50)(0.1)(0.9)} = \underline{2.12}$

Commands needed: CDF C3 C4;
                  NORMal 5 2.12.
with C3 containing -0.05 and 6.5.

$P(x \leq 6)$ = 0.760387 - 0.008608 = <u>0.751779</u>

6.99 a. $P[x \leq 75 | B(n = 300, p = 0.2)]$ = P(0) + P(1) + P(2) + P(3) +
                                              P(4) + ... + P(75)

b. CDF 75;
    BINOmial 300 0.2.   Result: <u>0.9856</u>

c. $\mu = np = (300)(0.2) = \underline{60}$
    $\sigma = \sqrt{npq} = \sqrt{(300)(0.2)(0.8)} = \underline{6.93}$

Commands needed: CDF C7 C8;
                  NORMal 60 6.93.
with C7 containing -0.05 and 75.5.

$P(x \leq 75)$ = 0.987346 - 0.0000 = <u>0.9873</u>

d. (b) and (c) result in answers that are very close in value.

6.101 $\mu = np = (100)(0.80) = \underline{80.0}$
$\sigma = \sqrt{npq} = \sqrt{(100)(0.80)(0.20)} = \underline{4.0}$

$$
\begin{aligned}
P(x \leq 70) &= P(x < 70.5) \\
&= P[z < (70.5 - 80.0)/4.0] \\
&= P[z < -2.38] \\
&= 0.5000 - 0.4913 = \underline{0.0087}
\end{aligned}
$$

6.103 a. $P(\text{exactly } 110 \text{ of } 200 \text{ agree}) = P[x = 110|B(200,0.58)]$

$$\mu = np = (200)(0.58) = \underline{116.0}$$

$$\sigma = \sqrt{npq} = \sqrt{(200)(0.58)(0.42)} = \underline{6.98}$$

$$
\begin{aligned}
P[x = 110] &= P[109.5 < x < 110.5] \\
&= P[(109.5 - 116)/6.98 < z < (110.5 - 116)/6.98] \\
&= P[-0.93 < z < -0.79] \\
&= 0.3238 - 0.2852 = \underline{0.0386}
\end{aligned}
$$

b. $
\begin{aligned}
P[x < 110] &= P[x < 109.5] \\
&= P[z < (109.5 - 116)/6.98] \\
&= P[z < -0.93] \\
&= 0.5000 - 0.3238 = \underline{0.1762}
\end{aligned}
$

c. $
\begin{aligned}
P[x > 100] &= P[x > 100.5] \\
&= P[z > (100.5 - 116)/6.98] \\
&= P[z > -2.22] \\
&= 0.5000 + 0.4868 = \underline{0.9868}
\end{aligned}
$

CHAPTER 7 ∇ SAMPLE VARIABILITY

Chapter Preview

Chapters 1 and 2 introduced the concept of a sample and its various measures. Measures of central tendency, measures of dispersion, and the shape of the distribution of the data give a single "snapshot" of the population from which the sample was taken. If repeated samples are taken and statistics noted, a clearer picture of the population from which the sample came will develop. These combined statistics will enable us to better predict the population's parameters. Chapter 7 works with this illustration of repeated sampling in the form of a sampling distribution. A sampling distribution is basically a probability distribution for a sample statistic. Therefore, there can be sampling distributions for the sample mean, for the sample standard deviation or for the sample range, to name a few. The significant results that will surface for the probability distribution of the mean specifically, will be contained in the Central Limit Theorem. This theorem justifies the use of the normal distribution in solving a wide range of problems.

SECTION 7.1 MARGIN EXERCISES

7.1 Each digit has the same chance of being selected; therefore, each sample combination has the same probability as any other combination.
$P(0) = 0.04$, since there is only one way out of the 25 samples to get an $\overline{x}$ of 0; i.e. $\{0,0\}$ and $1/25 = 0.04$.
$P(2) = 0.12$, since there are 3 ways out of the 25 samples to get an $\overline{x}$ of 2; i.e., $\{0,4\},\{2,2\},\{4,0\}$ and $3/25 = 0.12$.

7.2

x	P(x)	xP(x)	$x^2P(x)$
1	0.2	0.2	0.2
2	0.2	0.4	0.8
3	0.2	0.6	1.8
4	0.2	0.8	3.2
5	0.2	1.0	5.0
Σ	1.0ck	3.0	11.0

$\mu = \Sigma xP(x) = \underline{3.0}$

$\sigma = \sqrt{\Sigma x^2 P(x) - (\Sigma xP(x))^2}$

$= \sqrt{11.0 - (3.0)^2}$

$= \sqrt{2} = \underline{1.41}$

7.3

Class	Freq.
1.8-2.2	3
2.2-2.6	5
2.6-3.0	6
3.0-3.4	6
3.4-3.8	5
3.8-4.2	4
4.2-4.6	1
Σ	30

7.4

$\overline{x}$	f	$\overline{x}f$	$\overline{x}^2 f$
2.0	3	6.0	12.00
2.2	3	6.6	14.52
2.4	2	4.8	11.52
2.6	1	2.6	6.76
2.8	5	14.0	39.20
3.0	4	12.0	36.00
3.2	2	6.4	20.48
3.4	3	10.2	34.68
3.6	2	7.2	25.92
3.8	4	15.2	57.76
4.4	1	4.4	19.36
Σ	30	89.4	278.20

$\overline{\overline{x}}$ = mean of the sample means

$\overline{\overline{x}} = \Sigma \overline{x} f / \Sigma f = 89.4/30 = \underline{2.98}$

$s_{\overline{x}}$ = standard deviation of the sample means

$$s_{\overline{x}} = \sqrt{\frac{\Sigma \overline{x}^2 f - (\Sigma \overline{x}f)^2 / \Sigma f}{\Sigma f - 1}} = \sqrt{\frac{278.2 - (89.4)^2 / 30}{29}}$$

$= 0.63756 = \underline{0.638}$

7.5 a.

Classes	Freq.
4.5-6.5	1
6.5-8.5	1
8.5-10.5	4
10.5-12.5	2
12.5-14.5	1
14.5-16.5	3
Σ	12

b. Samples are not all the same size; the airlines have different size fleets.

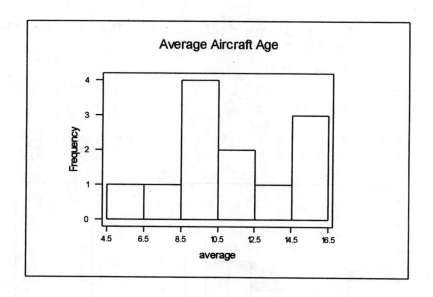

Average Aircraft Age

SECTION 7.1 EXERCISES

Use a tree diagram to find all possible samples for exercise 7.7a.
Each sample will have a probability of 1/n, where n is the number
of samples. Remember $\sum P(statistic) = 1$.

7.7 a.

11	31	51	71	91
13	33	53	73	93
15	35	55	75	95
17	37	57	77	97
19	39	59	79	99

b.

$\overline{x}$	1	2	3	4	5	6	7	8	9
$P(\overline{x})$	0.04	0.08	0.12	0.16	0.20	0.16	0.12	0.08	0.04

c.

R	0	2	4	6	8
$P(R)$	0.20	0.32	0.24	0.16	0.08

7.9 Every student will have different results, however they should be at least similar to these.

samples obtained:

| 899 | 091 | 031 | 982 | 159 | 612 | 720 | 534 | 758 | 178 |
| 337 | 520 | 185 | 893 | 601 | 968 | 560 | 959 | 368 | 943 |

a. 8.7 3.3 1.3 6.3 5.0 3.0 3.0 4.0 6.7 5.3
   4.3 2.3 4.7 6.7 2.3 7.7 3.7 7.7 5.7 5.3

b.

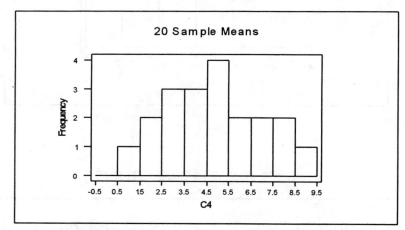

c. The shape is somewhat symmetric, centered around the 4.5 to 5.5 class.

d.

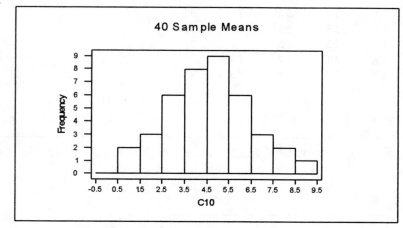

The shape is appearing more like a normal distribution and centered near 4.5.

MINITAB can do the work of repeated sampling from various
distributions. In exercise 7.11, the MINITAB commands simulate
sampling from a uniform distribution.

        RANDom 100 C1-C5;
        INTEger 0 to 9.
                - MINITAB randomly generates integers between the
                  values of 0 and 9 inclusive. Each value has the
                  same probability of being generated.
                  100 generated data values are put into each of
                  the columns C1, C2, C3, C4, and C5. We will
                  treat them as 100 samples of size 5.
                  (100 rows of 5 numbers)
        RMEAn C1-C5 PUT INTO C6
                - Means are calculated row-wise for the data in
                  columns 1 through 5. The 100 calculated sample
                  means, for the 100 samples of size 5, are put
                  into C6.

7.11 a. Every student will have different results, however the
             means of the 100 samples, each of size 5, should resemble
             those listed in (b).

       b. sample means

5.6	5.2	5.0	4.0	4.6	4.4	6.2	4.0	3.6	1.6	6.2
2.6	2.8	3.2	5.0	6.6	5.4	3.0	3.8	4.8	5.8	1.4
5.0	2.8	2.8	5.2	3.6	4.4	4.2	5.2	7.2	1.8	6.8
4.6	5.8	3.2	3.2	5.6	6.2	6.0	2.0	3.8	5.6	4.6
5.8	2.8	3.6	5.0	2.4	5.8	6.6	4.2	4.8	2.6	5.0
3.2	1.2	4.4	6.0	3.0	4.8	5.4	5.6	3.0	2.6	5.2
4.8	2.4	4.2	5.0	5.2	4.0	2.8	3.8	4.4	4.2	3.4
4.4	6.2	5.2	5.0	5.0	3.8	5.0	6.0	5.6	3.0	6.4
3.8	5.8	5.8	7.2	4.4	6.8	5.6	4.8	4.8	4.0	3.6
7.4										

c.

100 Sample Means, n=5

d. The shape is approximately normal, being mounded, approximately symmetrical, and centered near 4.5.

SECTION 7.2 MARGIN EXERCISES

7.13 $25/\sqrt{16}$ = <u>6.25</u> $25/\sqrt{36}$ = <u>4.167</u> $25/\sqrt{100}$ = <u>2.50</u>

7.14 a. one

b. $\sigma_{\bar{x}} = \sigma / \sqrt{n}$; as n increases the value of this fraction gets smaller.

The most important sampling distribution is the sampling distribution of the sample means. It provides the information that makes up the central limit theorem.

CENTRAL LIMIT THEOREM

If all possible random samples, each of size n, are taken from any population with a mean μ and standard deviation σ, the sampling distribution of sample means $\overline{x}$'s will result in the following:

1. The mean of the sample means (x bars) will equal the mean of the population, $\mu_{\overline{x}} = \mu$.

2. The standard deviation of the sample means (x bars) will be equal to the population standard deviation divided by the square root of the sample size, $\sigma_{\overline{x}} = \sigma / \sqrt{n}$.

3. A normal distribution when the parent population is normally distributed or becomes approximately normal distributed as the sample size increases when the parent population is not normally distributed.

In essence, $\overline{x}$ is normally distributed when n is large enough, no matter what shape the population is. The further the population shape is from normal, the larger the sample size needs to be.

$\sigma_{\overline{x}}$ = the standard deviation of the $\overline{x}$'s is now referred to as the <u>standard error of the mean</u>.

7.15 a. <u>500</u>   b. $30/\sqrt{36} = \underline{5}$ c. approximately normal

7.17 a. approximately normal   b. <u>$31.65</u>

   c. $12.25/\sqrt{150} = \underline{\$1.00}$

7.19 a. Every student will have different results; however, the means of the 100 samples, each of size 6, should resemble those listed in (b).

b. sample means

17.0148	21.0960	18.8767	18.4458	23.7957	17.4572	19.9137
22.9338	21.0164	18.9116	15.8072	23.1245	20.1439	20.8047
18.7836	19.5104	16.7224	18.1819	19.7173	19.4121	19.7335
18.9738	20.0119	19.7525	18.8456	19.3583	21.3249	20.7735
20.7584	18.9180	19.1935	22.2808	21.8498	19.2428	23.8004
19.1168	20.8396	22.2579	20.6043	17.9834	18.9368	20.2062
22.0984	19.4569	21.1661	19.9014	20.5755	17.0555	23.0511
21.3578	20.1454	19.3213	20.1349	18.3778	17.4479	21.5321
22.9323	18.8017	21.7476	18.1265	20.4839	20.3481	18.8614
21.2243	20.4734	17.4328	18.1990	23.0886	17.1433	18.4057
20.8029	20.7676	18.8139	18.2665	20.3459	20.8185	18.1914
19.1677	24.5807	22.6314	20.0144	17.6009	18.1102	19.1101
17.4663	17.1651	20.8377	22.2157	21.4562	19.4697	23.6880
18.8945	19.4205	19.7740	17.9856	20.2497	19.3615	21.2845
21.2142	17.3701					

c.

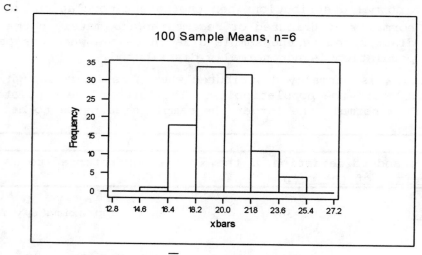

100 Sample Means, n=6

Mean of xbars $= \overline{\overline{x}} = 19.924$
Standard deviation of xbars $= s_{\overline{x}} = 1.8023$

d. $\overline{\overline{x}}$ is approximately 20.
$s_{\overline{x}}$ is approximately $4.5/\sqrt{6} = 1.84$.
The shape of the histogram is approximately normal.

SECTION 7.3 MARGIN EXERCISES

7.21 Both 90 and 110 are two standard errors away from the mean and from Table 3, the $P(0 < z < 2) = 0.4772$.

7.22 $P(\overline{x} < 39.75) = P[z < (39.75 - 39.0)/(2/\sqrt{25})]$
$$= P[z < 1.88]$$
$$= 0.5000 + 0.4699 = \underline{0.9699}$$

7.23 If $z = -0.67$, then
$$-0.67 = (\overline{x} - 39.0)/(2/\sqrt{25})$$
$$-0.268 = \overline{x} - 39.0$$
$$\overline{x} = 38.732 = \underline{38.73 \text{ inches}}$$

SECTION 7.3 EXERCISES

It is helpful to draw a normal curve, locating μ and shading in the desired portion for each problem. A new z formula must now be used to determine probabilities about $\overline{x}$.

$$z = \frac{\overline{x} - \mu}{\sigma / \sqrt{n}}$$

7.25 a. Heights are approximately normally distributed with a $\mu = 69$ and $\sigma = 4$.

b. $P(x > 70) = P[z > (70 - 69)/4]$
$$= P[z > 0.25]$$
$$= 0.5000 - 0.0987 = \underline{0.4013}$$

c. The distribution of $\overline{x}$'s will be approximately normally distributed.

d. $\mu_{\overline{x}} = \underline{69}$; $\sigma_{\overline{x}} = 4/\sqrt{16} = \underline{1.0}$

e. $P(\overline{x} > 70) = P[z > (70 - 69)/1.0]$
$$= P[z > +1.00]$$
$$= 0.5000 - 0.3413 = \underline{0.1587}$$

f. $P(\bar{x} < 67) = P[z < (67 - 69)/1.0]$
$= P[z < -2.00]$
$= 0.5000 - 0.4772 = \underline{0.0228}$

Watch the wording of the various probability problems.

If the probability for an individual item or person (x) is desired, use $z = (x - \mu)/\sigma$.

If the probability for a sample mean $\bar{x}$ is desired, use $z = (\bar{x} - \mu) / (\sigma / \sqrt{n})$.

7.27 a. $P(38 < x < 40) = P[(38 - 39)/2 < z < (40 - 39)/2]$
$= P[-0.50 < z < +0.50]$
$= 0.1915 + 0.1915 = \underline{0.3830}$

b. $P(38 < \bar{x} < 40) =$
$= P[(38 - 39)/(2/\sqrt{30}) < z < (40 - 39)/(2/\sqrt{30})]$
$= P[-2.74 < z < +2.74]$
$= 0.4969 + 0.4969 = \underline{0.9938}$

c. $P(x > 40) = P[z > (40 - 39)/2]$
$= P[z > 0.50]$
$= 0.5000 - 0.1915 = \underline{0.3085}$

d. $P(\bar{x} > 40) = P[z > (40 - 39)/(2/\sqrt{30})]$
$= P[z > 2.74]$
$= 0.5000 - 0.4969 = \underline{0.0031}$

7.29 a. $P(375.00 < \bar{x} < 385.00)$
$= P[(375 - 381)/(85/\sqrt{250}) < z < (385 - 381)/(85/\sqrt{250})]$
$= P[-1.12 < z < 0.74]$
$= 0.3686 + 0.2704 = \underline{0.6390}$

b. Since the sample size is n = 250, the central limit theorem holds and the use of the normal distribution is justifiable in answering questions about sample means.

c. If the distribution is normal, the median is equal or approximately equal to the mean.

d. Probably not; salaries typically have a skewed distribution.

e. It probably increased the probability since $381 is within the interval. If the actual mean is higher than $385, then the actual probability will be less than the calculated value.

7.31 a. Using MINITAB: $P(4 < \bar{x} < 6) = CDF(6) - CDF(4)$
$$= 0.841345 - 0.158655 = \underline{0.68269}$$

Using Table 3: $P(4 < \bar{x} < 6) =$
$$= P[(4 - 5)/(2/\sqrt{4}) < z < (6 - 5)/(2/\sqrt{4})]$$
$$= P[-1.00 < z < +1.00]$$
$$= 0.3413 + 0.3413 = \underline{0.6826}$$

Everybody will get different answers, however, the results should be similar to the results below.

b. sample means

4.45967	4.04628	4.56959	4.74174	5.00541	4.99137	5.27322
4.18889	6.16806	5.10562	4.21561	5.61553	4.73023	3.23896
5.14986	5.83925	3.87701	4.15342	6.95931	6.69219	4.92261
4.00410	5.33392	5.69277	4.04213	6.00010	5.13663	7.04077
4.95368	6.07187	4.61556	5.96161	5.95970	5.93136	4.20692
6.13040	3.54086	4.73198	4.18744	5.05378	3.01671	3.47408
4.93583	4.18053	6.02229	3.94990	4.81555	5.11444	3.43246
4.81337	5.51779	3.50610	3.70504	5.49885	4.25768	4.01218
6.07272	4.52822	5.57177	4.35903	3.49724	4.20362	4.52965
5.47631	4.86655	3.91419	4.71412	5.39523	4.63979	5.28885
3.65115	4.72093	3.48571	3.73255	5.33954	3.99134	4.62782
6.29056	4.51388	5.25607	6.30200	6.23775	5.14931	3.57207
6.28936	4.45702	4.57191	5.31639	6.45744	3.70814	5.61942
4.47878	3.06674	5.69816	4.75256	5.66117	4.08780	3.88444
4.93384	5.66094					

c.

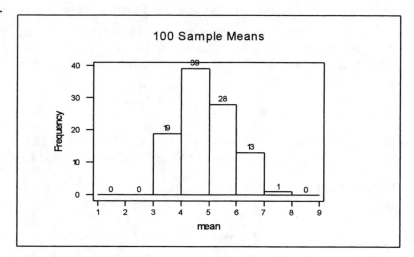

100 Sample Means

Inspecting the histogram, we find <u>67</u> of the sample means
are between 4 and 6;   67/100 = 67%

d. 67% is very close to the expected 0.6826.

CHAPTER EXERCISES

7.33 a. P(x < 2.25) = P[z < (2.25 - 2.63)/0.25]
                    = P[z < -1.52]
                    = 0.5000 - 0.4357 = <u>0.0643</u> or <u>6.43%</u>

b. P(x > 2.56) = P[z > (2.56 - 2.63)/0.25]
                    = P[z > -0.28]
                    = 0.5000 + 0.1103 = <u>0.6103</u> or <u>61.03%</u>

c. P($\overline{x}$ > 2.56) = P[z > (2.56 - 2.63)/(0.25/$\sqrt{100}$]
                    = P[z > -2.80]
                    = 0.5000 + 0.4974 = <u>0.9974</u>

d. z is used in (a) and (b) since the distribution of x is
    given to be normal, and z is used in (c) since the
    sampling distribution of x-bar is normal.

e. (a) and (b) are distributions of individual x-values,
    while (c) is a sampling distribution of $\overline{x}$-values.

7.35 $P(60 < \overline{x} < 70)$

$\qquad = P[(60 - 64.9)/(17.5/\sqrt{100}) < z < (70 - 64.9)/(17.5/\sqrt{100})]$

$\qquad = P[-2.80 < z < +2.91]$

$\qquad = 0.4974 + 0.4982 = \underline{0.9956}$

7.37 a. $P(245 < x < 255)$

$\qquad = P[(245 - 235)/\sqrt{400} < z < (255 - 235)/\sqrt{400}]$

$\qquad = P[+0.50 < z < +1.00]$

$\qquad = 0.3413 - 0.1915 = \underline{0.1498}$

b. $P(\overline{x} > 250) = P[z > (250 - 235)/(20/\sqrt{10})]$

$\qquad\qquad\qquad = P[z > +2.37]$

$\qquad\qquad\qquad = 0.5000 - 0.4911 = \underline{0.0089}$

7.39 $P(425 < \overline{x} < 475)$

$\qquad = P[(425 - 450)/(125/\sqrt{100}) < z < (475 - 450)/(125/\sqrt{100})]$

$\qquad = P[-2.00 < z < +2.00]$

$\qquad = 0.4772 + 0.4772 = \underline{0.9544}$

7.41 $P(\overline{x} < 680) = P[z < (680 - 700)/(120/\sqrt{144})]$

$\qquad\qquad\qquad = P[z < -2.00]$

$\qquad\qquad\qquad = 0.5000 - 0.4772 = \underline{0.0228}$

7.43 $P(\Sigma x > 38,000) = P[z > (\Sigma x - n\mu)/(\sigma\sqrt{n})]$

$\qquad\qquad\qquad = P[z > (38,000 - (50)(750))/(25\sqrt{50})]$

$\qquad\qquad\qquad = P[z > 500/176.777]$

$\qquad\qquad\qquad = P[z > +2.83]$

$\qquad\qquad\qquad = 0.5000 - 0.4977 = \underline{0.0023}$

7.45 a. Let Σx represent the total weight:

$\qquad P(\Sigma x > 4000) = P(\Sigma x/n > 4000/25)$

$\qquad\qquad\qquad = P(\overline{x} > 160)$

$\qquad\qquad\qquad = P[z > (160 - 300)/(50/\sqrt{25})]$

$\qquad\qquad\qquad = P[z > -14.0]$

$\qquad\qquad\qquad = \text{approximately } \underline{1.000}$

b. $P(\sum x < 8000) = P(\overline{x} < 320)$
$$= P[z < (320 - 300)/(50/\sqrt{25})]$$
$$= P[z < 2.00]$$
$$= 0.5000 + 0.4772 = \underline{0.9772}$$

For exercise 7.47, use $z = \dfrac{\overline{x} - \mu}{\sigma / \sqrt{n}}$. Substitute given values into the equation and solve for the unknown. A picture diagram is usually very helpful in organizing all the information.

7.47 a. $P(\overline{x} > 140.0) = P[z > (140.0 - 135.0)/(10.0/\sqrt{10})]$
$$= P[z > +1.58]$$
$$= 0.5000 - 0.4429 = \underline{0.0571}$$

b.

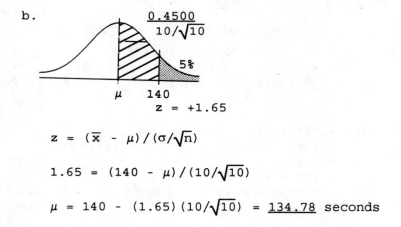

$$z = (\overline{x} - \mu)/(\sigma/\sqrt{n})$$

$$1.65 = (140 - \mu)/(10/\sqrt{10})$$

$$\mu = 140 - (1.65)(10/\sqrt{10}) = \underline{134.78} \text{ seconds}$$

7.49 a. $\mu = np = 16(0.5) = \underline{8.0}$
$$\sigma = \sqrt{npq} = \sqrt{16(0.5)(0.5)} = \sqrt{4} = \underline{2}$$

b.

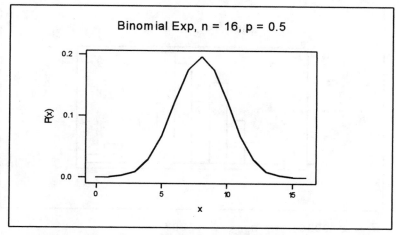

Binomial Exp, n = 16, p = 0.5

The commands given in JES2 produce a graph of the
distribution, as shown above. If a table of values is
desired, use the following MINITAB commands:

```
SET C1
0:16
END
PDF C1 C2;
BINO 16 0.5.
```

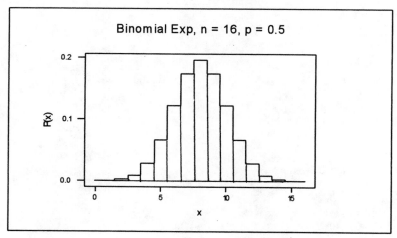

Binomial Exp, n = 16, p = 0.5

c. Every student will have different results, however the
histograms should resemble that in (d).

d.

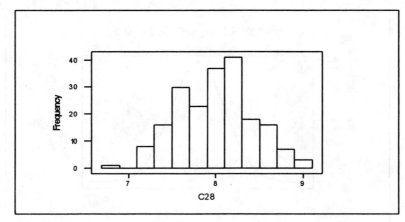

$\overline{\overline{x}} = 8.0036, \quad s_{\overline{x}} = 0.42287$

e. The mean of the sample means, $\overline{\overline{x}} = 8.0036$, is
   approximately equal to $\mu = 8$. The standard deviation of
   the sample means, $s_{\overline{x}} = 0.42287$, is approximately equal to
   $2/\sqrt{25} = 0.4$. The distributions are both approximately
   normally distributed.

CHAPTER 8 ▽ INTRODUCTION TO STATISTICAL

Chapter Preview

Chapter 8 introduces inferential statistics. Generalizations about population parameters are made based on sample data in inferential statistics. These generalizations can be made in the form of hypothesis tests or confidence interval estimations. Each is calculated with a degree of uncertainty. The integral elements and procedure for obtaining a confidence interval and for completing a hypothesis test will be presented in this chapter. They will be performed with respect to the population mean, μ. The population standard deviation, σ, will be considered as a known quantity.

SECTION 8.1 MARGIN EXERCISES

8.1 Difficulty and collector fatigue in obtaining and also in evaluating a very large sample; cost of sampling; destruction of product in cases like the rivets illustration.

8.2 $\sigma_{\bar{x}} = \sigma / \sqrt{n} = 18 / \sqrt{36} = 18 / 6 = 3$

8.3
$$P(\mu - 2\sigma_{\bar{x}} < \bar{x} < \mu + 2\sigma_{\bar{x}}) = P(-2\sigma_{\bar{x}} < \bar{x} - \mu < +2\sigma_{\bar{x}})$$
$$= P(-2 < (\bar{x} - \mu)/\sigma_{\bar{x}} < +2)$$
$$= P(-2 < z < 2)$$
$$= 0.4772 + 0.4772 = 0.9544 = \underline{95.44\%}$$

8.4 a. 33
   b. 95% of the time the parameter being estimated will be a value within the resulting interval. The data in this case study estimates the parameter to be 33, or within the interval between 30 and 36.
   c. 52 is the point estimate for μ. ± 5 is one-half of the interval width which is then 10 units. 52 ± 5 represents the interval with bounds from 47 to 57.

ESTIMATION

Point Estimate for a Population Parameter - the value of the
corresponding sample
statistic

    ex.) $\overline{x}$ is the point estimate of μ
        s is the point estimate of σ
        s^2 is the point estimate of σ^2

Confidence intervals are used to estimate a population parameter on
an interval with a degree of certainty. We could begin by taking a
sample and finding $\overline{x}$ to just estimate μ. The sample statistic, $\overline{x}$,
is a <u>point estimate</u> of the population parameter μ. How good an
estimate it is depends not only on the sample size and variability
of the data, but also whether or not the sample statistic is
unbiased.

Unbiased Statistic - a sample statistic whose sampling distribution
has a mean value equal to the corresponding
population parameter

One would assume that $\overline{x}$ is not exactly equal to μ, but hopefully
relatively close. It is by this reasoning that we work with
interval estimates of population parameters.

Level of Confidence = $1 - \alpha$ = the probability that the interval
constructed, based on the sample, contains the true population
parameter.

8.5   a. 24 = sample size = n;   4'11" = sample mean = $\overline{x}$
       b. 16 = population standard deviation = σ
       c. 190 = sample variance = s^2
       d. 69 = population mean = μ

8.7   a. II has the lower variability; both have a mean value equal
          to the parameter.
       b. II has a mean value equal to the parameter, I does not.
       c. Neither is a good choice; I is negatively biased with less
          variability, while II is only slightly positively biased
          with a larger variability.

8.9 $\mu_{\bar{x}} = \mu$; $\sigma_{\bar{x}} = \sigma / \sqrt{n}$ decreases as n increases

8.11   48($1)   = $ 48
           27($5)   =   135
           16($10) =   160
            6($25) =   150
            3($50) =   150
                        $643 per 100 cards;
         therefore 250($643) = $160,750

SECTION 8.2 MARGIN EXERCISES

8.13   The point estimate is the center of the confidence interval. The maximum error is based on the level of confidence, sample size and standard deviation. The confidence interval is bounded by the two numbers found by adding and subtracting the point estimate and the maximum error.

8.14   a. 75.92

   b. $E = z(\alpha/2) \cdot \sigma/\sqrt{n} = (2.33)(0.5/\sqrt{10}) = 0.3684 = 0.368$

   c. $\bar{x} \pm E = 75.92 \pm 0.368$

            75.552 to 76.288, the 0.98 confidence interval for μ

8.15   The numbers are calculated for samples of snow for some of the months; it would be impossible to gather this information from every inch of snow.

8.16   Each student will have different results; however, 4.5 should be in the interval about 90% of the time.

8.17 $n = [z(\alpha/2) \cdot \sigma/E]^2 = [(2.33)(3)/1]^2 = 48.8601 = 49$

SECTION 8.2 EXERCISES

ESTIMATION OF THE POPULATION MEAN - μ

Point estimate of μ: $\overline{x}$

Interval estimate of μ = confidence interval

$1 - \alpha$ = level of confidence, the probability or degree of
certainty desired (ex.: 95%, 99% ...)

A $(1-\alpha)$ confidence interval estimate for μ is:

$$\overline{x} - z(\alpha/2) \cdot \sigma/\sqrt{n} \quad \text{to} \quad \overline{x} + z(\alpha/2) \cdot \sigma/\sqrt{n} \quad **$$

$\overline{x} - z(\alpha/2) \cdot \sigma/\sqrt{n}$ = lower confidence limit

$\overline{x} + z(\alpha/2) \cdot \sigma/\sqrt{n}$ = upper confidence limit

$E = z(\alpha/2) \cdot \sigma/\sqrt{n}$ = maximum error of the estimate

**To find $z(\alpha/2)$:
    (suppose for example that a 95% confidence interval is desired)
                95% = $1 - \alpha$, that is,
                .95 = $1 - \alpha$
           solving for alpha, α, gives
                α = .05
          dividing both sides by 2 gives
             $\alpha/2$ = .025
Now determine the probability associated with $z(.025)$ using Table
4B (Appendix B, JES2-p522), the Critical Values of Standard Normal
Distribution for Two-Tailed Situations. (This table conveniently
gives the most popular critical values for z.)

8.19 Either the sampled population is normally distributed or the
random sample is sufficiently large for the central limit
theorem to hold.

```
┌─────────────────────────────────────────────────────────────────────┐
│          THE CONFIDENCE INTERVAL: A FOUR-STEP PROCEDURE             │
│                                                                     │
│  Step 1: Describe the population parameter of concern.             │
│  Step 2: Specify the confidence interval criteria.                 │
│          a. Check the assumptions.                                  │
│          b. Determine the test statistic to be used.               │
│          c. State the level of confidence, 1 - α.                  │
│  Step 3: Collect and present sample evidence.                      │
│          a. Collect the sample information.                         │
│          b. Find the point estimate.                                │
│  Step 4: Determine the confidence interval.                        │
│          a. Determine the confidence coefficients.                 │
│          b. Find the maximum error or estimate.                    │
│          c. Find the lower and upper confidence limits.            │
│          d. Describe the results.                                   │
└─────────────────────────────────────────────────────────────────────┘
```

8.21 a. Step 1: The mean, μ

Step 2: a. normality indicated

b. z, $\sigma = 6$ c. $1-\alpha = 0.95$

Step 3: a. $n = 16$, $\overline{x} = 28.7$

b. $\overline{x} = 28.7$

Step 4: a. $\alpha/2 = 0.05/2 = 0.025$; $z(0.025) = 1.96$

b. $E = z(\alpha/2) \cdot \sigma/\sqrt{n} = (1.96)(6/\sqrt{16})$

$= (1.96)(1.5) = 2.94$

c. $\overline{x} \pm E = 28.7 \pm 2.94$

<u>25.76 to 31.64</u>, the 0.95 confidence interval for μ

b. <u>Yes</u>; the sampled population is normally distributed.

8.23 a. The mean amount spent for textbooks per student during the fall semester at a large community college.

b. Step 1: See part 'a'

Step 2: a. normality assumed, CLT with $n = 75$.

b. z, $\sigma = 35$ c. $1-\alpha = 0.90$

Step 3: a. $n = 75$, $\overline{x} = 158.30$

b. $\overline{x} = 158.30$

Step 4: a. $\alpha/2 = 0.10/2 = 0.05$; $z(0.05) = 1.65$

b. $E = z(\alpha/2) \cdot \sigma/\sqrt{n} = (1.65)(35/\sqrt{75})$

$= (1.65)(4.04) = 6.67$

c. $\overline{x} \pm E = 158.30 \pm 6.67$

<u>151.63 to 164.97</u>, the 0.90 confidence interval for μ

REMEMBER: variance = σ^2 , therefore $\sigma = \sqrt{\sigma^2}$

8.25   a. 25.3

b. Step 1: The mean age of night school students
Step 2: a. normality assumed, CLT with n = 60.
   b. z, σ^2 = 16 or σ = 4   c. 1-α = 0.95
Step 3: a. n = 60, $\overline{x}$ = 25.3
   b. $\overline{x}$ = 25.3
Step 4: a. $\alpha/2$ = 0.05/2 = 0.025; z(0.025) = 1.96
   b. E = z($\alpha/2$)·$\sigma/\sqrt{n}$ = (1.96)(4/$\sqrt{60}$)
         = (1.96)(0.516) = 1.01
   c. $\overline{x}$ ± E = 25.3 ± 1.01
   <u>24.29 to 26.31</u>, the 0.95 confidence interval for μ

c. Step 1-3: as shown in (b), except 2c. 1-α = 0.99
Step 4: a. $\alpha/2$ = 0.01/2 = 0.005; z(0.005) = 2.58
   b. E = z($\alpha/2$)·$\sigma/\sqrt{n}$ = (2.58)(4/$\sqrt{60}$)
         = (2.58)(0.516) = 1.33
   c. $\overline{x}$ ± E = 25.3 ± 1.33
   <u>23.97 to 26.63</u>, the 0.99 confidence interval for μ

8.27   a. <u>55.20</u>

b. E = z($\alpha/2$)·$\sigma/\sqrt{n}$ = (1.96)(19.5/$\sqrt{20}$) = <u>8.546</u>

c. $\overline{x}$ ± E = 55.20 ± 8.546
   <u>46.654</u> <u>to</u> <u>63.746</u> is the 0.95 confidence interval for μ

d. They are the main parts of the 1-α confidence interval.

To find the <u>sample size n</u> required for a 1-α confidence interval,
use the formula: n = [z($\alpha/2$)·σ/E]², where

   z = standard normal distribution
   α = calculated from the 1-α confidence interval desired
   σ = population standard deviation
   E = maximum error of the estimate

The maximum error of the estimate, E, is the amount of error that
is tolerable or allowed. Quite often, finding the word "within" in
an exercise will locate the acceptable value for E.

8.29   n = [z($\alpha/2$)·σ/E]² = [(2.58)(900)/75]² = 958.5 = <u>959</u>

8.31 a. $n = [z(\alpha/2) \cdot \sigma/E]^2 = [(1.96)(5)/1]^2 = 96.04 = \underline{97}$

b. $n = [z(\alpha/2) \cdot \sigma/E]^2 = [(1.96)(5)/2]^2 = 24.01 = \underline{25}$

SECTION 8.3 MARGIN EXERCISES

8.33 H_o: The system is reliable
H_a: The system is not reliable

8.34 H_a: Teaching techniques have a significant effect on student's exam scores.

8.35 Type A correct decision:
Truth of situation: the party will be a dud.
Conclusion: the party will be a dud.
Action: did not go [avoided dud party]

Type B correct decision:
Truth of situation: the party will be a great time.
Conclusion: the party will be a great time.
Action: did go [party was great time]

Type I error:
Truth of situation: the party will be a dud.
Conclusion: the party will be a great time.
Action: did go [party was a dud]

Type II error:
Truth of situation: the party will be a great time.
Conclusion: the party will be a dud.
Action: did not go [missed great party]

Remember; the truth of the situation is not known before the decision is made, the conclusion reached and the resulting actions take place. Only after the party is over can the evaluation be made.

8.36 You missed a great time.

8.37 α is the probability of rejecting a TRUE null hypothesis; $1-\beta$ is the probability of rejecting a FALSE null hypothesis; they are two distinctly different acts that both result in rejecting the null hypothesis.

8.38 The smaller the probability of an event is, the less often it occurs.

SECTION 8.3 EXERCISES

DEFINITIONS FOR HYPOTHESIS TESTS

Hypothesis - a statement that something is true.

Null Hypothesis, H_O - a statement that specifies a value for a population parameter
ex.: H_O: The mean weight is 40 pounds

Alternative Hypothesis, H_a - opposite of H_O, a statement that specifies an "opposite" value for a population parameter
ex.: H_a: The mean weight is not 40 pounds

Type I Error - the error resulting from rejecting a true null

α (alpha) - the probability of a type I error, that is the probability of rejecting H_O when it is true.

Type II Error - the error resulting from not rejecting a false null hypothesis

β (beta) - the probability of a type II error, that is the probability of not rejecting H_O when it is false.

Keep α and β as small as possible, depending on the severity of the respective error.

8.39 a. H_O: Special delivery mail does not take too much time
H_a: Special delivery mail takes too much time

b. H_O: The new design is not more comfortable
   H_a: The new design is more comfortable

c. H_O: Cigarette smoke has no effect on the quality of a
   person's life
   H_a: Cigarette smoke has an effect on the quality of a
   person's life

d. H_O: The hair conditioner is not effective on "split ends"
   H_a: The hair conditioner is effective on "split ends"

8.41 a. H_O: The victim is alive
   H_a: The victim is not alive

b. Type A correct decision: The victim _is_ _alive_ and is
   treated _as_ _though_ _alive_.

   Type I error: The victim _is_ _alive_, but is treated
   _as_ _though_ _dead_.

   Type II error: The victim _is_ _dead_, but treated _as_
   if alive.

   Type B correct decision: The victim _is_ _dead_ and treated
   _as_ _dead_.

c. The type I error is very serious. The victim may very
   well be dead shortly without the attention that is not
   being received.

   The type II error is not as serious. The victim is
   receiving attention that is of no value. This would be
   serious only if there were other victims that needed this
   attention.

8.43 a. A type I error occurs when it is determined that the
   majority of Americans do not favor laws against assault
   weapons when, in fact, the majority do favor such laws.
   A type II error occurs when it is determined that the
   majority of Americans do favor laws against assault
   weapons when, in fact, they do not favor such laws.

b. A type I error occurs when it is determined that the fast food is low salt when, in fact, it is not low salt.
A type II error occurs when it is determined that the fast food is not low salt when, in fact, it is low salt.

c. A type I error occurs when it is determined that the building must be demolished when, in fact, it should not be demolished.
A type II error occurs when it is determined that the building must not be demolished when, in fact, it should be demolished.

d. A type I error occurs when it is determined that there is waste in government spending when, in fact, there is not waste.
A type II error occurs when it is determined that there is no waste in government spending when, in fact, there is waste.

TERMINOLOGY FOR DECISIONS IN HYPOTHESIS TESTS

Reject H_O: use when the evidence disagrees with the null hypothesis.

Fail to reject H_O: use when the evidence does not disagree with the null hypothesis.

Note: The purpose of the hypothesis test is to allow the evidence a chance to discredit the null hypothesis.
Remember: If one believes the null hypothesis to be true, generally there is no test.

8.45 a. Type I   b. Type II   c. Type I   d. Type II

8.47 a. The type I error is very serious and, therefore, we are willing to allow it to occur with a probability of 0.001; that is, only 1 chance in 1000.

b. The type I error is somewhat serious and, therefore, we are willing to allow it to occur with a probability of 0.05; that is, 1 chance in 20.

c. The type I error is not at all serious and, therefore, we are willing to allow it to occur with a probability of 0.10; that is, 1 chance in 10.

8.49 a. α b. β

8.51 a. When the test procedure begins, the experimenter is thoroughly convinced the alternative hypothesis can be shown to be true; thus when the decision *reject H_O* is attained, the experimenter will want to say something like "see I told you so." Thus the statement of the conclusion is a fairly strong statement like; "the evidence shows beyond a shadow of a doubt (is significant) that the alternative hypothesis is correct."

b. When the test procedure begins, the experimenter is thoroughly convinced the alternative hypothesis can be shown to be true; thus when the decision *fail to reject H_O* is attained, the experimenter is disappointed and will want to say something like "okay so this evidence was not significant, I'll try again tomorrow." Thus the statement of the conclusion is a fairly mild statement like, "the evidence was not sufficient to show the alternative hypothesis to be correct."

8.53 a. α = P(rejecting H_O when the H_O is true)

$= P(x \geq 86 | \mu=80) = P(z > (86 - 80)/5) = P(z > 1.20)$
$= 0.5000 - 0.3849 = \underline{0.1151}$

b. β = P(accepting H_O when the H_O is false)

$= P(x < 86 | \mu=90) = P(z < (86 - 90)/5) = P(z < -0.80)$
$= 0.5000 - 0.2881 = \underline{0.2119}$

SECTION 8.4 MARGIN EXERCISES

8.55 H_O: The mean shearing strength is at least 925 lbs.
H_a: The mean shearing strength is less than 925 lbs.

8.56 H_O: μ = 54.4
H_a: $\mu \neq 54.4$

8.57 a. H_O: $\mu = 1.25$ ($\leq$)
   H_a: $\mu > 1.25$

b. H_O: $\mu = 335$ ($\geq$)
   H_a: $\mu < 335$

c. H_O: $\mu = 230,000$
   H_a: $\mu \neq 230,000$

8.58 Type A correct decision: The mean shearing strength is at least 925 lbs and it is decided that it is.
Type I error: The mean shearing strength is at least 925 lbs and it is decided that it is less than 925 lbs.
Type II error: The mean shearing strength is less than 925 lbs and it is decided that it is greater than or equal to 925 lbs.
Type B correct decision: The mean shearing strength is less than 925 lbs and it is decided that it is less than 925 lbs.

Type II error; you buy and use weak rivets.

8.59 $z = (\overline{x} - \mu)/(\sigma/\sqrt{n})$
$z* = (54.3 - 56)/(7/\sqrt{36}) = \underline{-1.46}$

8.60 a. p-value = $P(z < -2.3) = P(z > 2.3)$
   $= 0.5000 - 0.4893 = \underline{0.0107}$

b. p-value = $P(z > 1.8) = 0.5000 - 0.4641 = \underline{0.0359}$

8.61 a. Fail to reject H_O b. Reject H_O

8.62 a. $z* = (\overline{x} - \mu)/(\sigma/\sqrt{n}) = (24.5 - 22.5)/(6/\sqrt{36}) = \underline{2.0}$
   p-value = $P(z > 2.0) = 0.5000 - 0.4772 = \underline{0.0228}$

b. $z* = (\overline{x} - \mu)/(\sigma/\sqrt{n}) = (192.5 - 200)/(40/\sqrt{50}) = \underline{-1.33}$
   p-value = $P(z < -1.33) = P(z > 1.33)$
   $= 0.5000 - 0.4082 = \underline{0.0918}$

c. $z* = (\overline{x} - \mu)/(\sigma/\sqrt{n}) = (11.52 - 12.4)/(2.2/\sqrt{16}) = \underline{-1.6}$

p-value $= 2 \cdot P(z < -1.6) = 2 \cdot P(z > 1.6)$
$= 2(0.5000 - 0.4452) = 2(0.0548) = \underline{0.1096}$

8.63 p-value $= 2 \cdot P(z > 1.1) = 2(0.5000 - 0.3643) = 2(0.1357)$
$= \underline{0.2714}$

8.64 The p-value measures the likeliness of the sample results based on a true null hypothesis.

8.65 Results will vary; however, expect your results to be similar to those shown in Table 8-7.

8.66 N = n and is the number of data values
MEAN calculated using formula $\Sigma x/n$
STDEV calculated using formula $\sqrt{\Sigma(x - \overline{x})^2/(n - 1)}$
SEMEAN calculated using formula $\sigma/\sqrt{n}$
z calculated using formula $(\overline{x} - \mu)/(\sigma/\sqrt{n})$
p-value calculated using formula $P(z < -1.50)$

SECTION 8.4 EXERCISES

THE PROBABILITY-VALUE HYPOTHESIS TEST: A FIVE-STEP PROCEDURE

Step 1: Describe the population parameter of concern.
Step 2: State the null hypothesis (H_O) and the alternative hypothesis (H_a).
Step 3: Specify the test criteria.
    a. Check the assumptions.
    b. Identify the test statistic to be used.
    c. Determine the level of significance, α.
Step 4: Collect and present the sample evidence.
    a. Collect the sample information.
    b. Calculate the value of the test statistic.
    c. Calculate the p-value.   ...

```
Step 5: Determine the results.
        a. Determine whether or not the p-value is smaller than α.
        b. Make a decision about H_O.
        c. Write a conclusion about H_a.
```

The Null and Alternative Hypotheses, H_O and H_a

H_O: $\mu = 100$ versus H_a: $\mu \neq 100$ (= and ≠ form the opposite of each other)

        H_a is a two-sided alternative
        possible wording for this combination:
                a) mean is different from 100 ($\neq$)
                b) mean is not 100 ($\neq$)
                c) mean is 100 (=)

    OR ───

H_O: $\mu = 100$ ($\leq$) versus H_a: $\mu > 100$ ($\leq$ and > form the opposite of each other)

        H_a is a one-sided alternative
        possible wording for this combination:
                a) mean is greater than 100 (>)
                b) mean is at most 100 ($\leq$)
                c) mean is no more than 100 ($\leq$)

    OR ───

H_O: $\mu = 100$ ($\geq$) versus H_a: $\mu < 100$ ($\geq$ and < form the opposite of each other)

        H_a is a one-sided alternative
        possible wording for this combination:
                a) mean is less than 100 (<)
                b) mean is at least 100 ($\geq$)
                c) mean is no less than 100 ($\geq$)

───

Always show equality (=) in the null hypothesis, since the null hypothesis must specify a single specific value for μ (like $\mu = 100$).
The null hypothesis could be rejected in favor of the alternative hypothesis for three different reasons; 1) $\mu \neq 100$ or 2) $\mu > 100$ or 3) $\mu < 100$. Together, the two opposing statements, H_O and H_a, must contain or account for all numerical values around and including μ. This allows for the addition of $\geq$ or $\leq$ to the null hypothesis.

...

Therefore, if the alternative hypothesis is < or >, ≥ or ≤, respectively, may be added to the null hypothesis. If ≥ or ≤ is being tested, the appropriate symbol should be written in parentheses after the amount stated for μ.
Sometimes, depending on the wording, it is easier to write the alternative hypothesis first. The alternative hypothesis can only contain >, < or ≠.

Hint: Sometimes it is helpful to either: 1) <u>remove</u> the word "no" or "not" when it is included in the claim, or 2) <u>insert</u> "no" or "not" when it is not in the claim, to form the opposite of the claim.

8.67 a. H_O: $\mu = 26$ yrs (≤) vs. H_a: $\mu > 26$

b. H_O: $\mu = 36.7$ lbs (≥) vs. H_a: $\mu < 36.7$

c. H_O: $\mu = 1600$ hrs (≥) vs. H_a: $\mu < 1600$

d. H_O: $\mu = 210$ lbs (≤) vs. H_a: $\mu > 210$

e. H_O: $\mu = 570$ lbs/unit vs. H_a: $\mu \neq 570$

Use the sample information (sample mean and size) and the population parameter in the null hypothesis (μ) to calculate the test statistic z^*.
$$z^* = (\overline{x} - \mu)/(\sigma/\sqrt{n})$$

8.69 a. $z = (\overline{x} - \mu)/(\sigma/\sqrt{n})$

$z^* = (10.6 - 10)/(3/\sqrt{40}) = \underline{1.26}$

b. $z = (\overline{x} - \mu)/(\sigma/\sqrt{n})$

$z^* = (126.2 - 120)/(23/\sqrt{25}) = \underline{1.35}$

c. $z = (\overline{x} - \mu)/(\sigma/\sqrt{n})$

$z^* = (18.93 - 18.2)/(3.7/\sqrt{140}) = \underline{2.33}$

d. $z = (\overline{x} - \mu)/(\sigma/\sqrt{n})$

$z^* = (79.6 - 81)/(13.3/\sqrt{50}) = \underline{-0.74}$

DECISIONS AND CONCLUSIONS

Since the null hypothesis, H_O, is usually thought to be the statement whose truth is being challenged by the experimenter, all decisions are about the null hypothesis. The alternative hypothesis, H_a, however is usually thought to express the experimenter's viewpoint. Thus, the interpretation of the decision or conclusion is expressed from the experimenter and alternative hypothesis point of view.

Decision:

1) If the p-value is less than or equal to the specified level of significance (α), the null hypothesis will be rejected (if $P \le \alpha$, reject H_O).

2) If the p-value is greater than the specified level of significance (α), fail to reject the null hypothesis (if $P > \alpha$, fail to reject H_O).

Conclusion:

1) If the decision is "reject H_O," the conclusion should read "There is sufficient evidence at the α level of significance to show that ... (the meaning of the alternative hypothesis)."

2) If the decision is "fail to reject H_O," the conclusion should read "There is not sufficient evidence at the α level of significance to show that ... (the meaning of the alternative hypothesis)."

8.71 a. *Reject H_O* or *Fail to reject H_O*

b. When the calculated p-value is smaller than or equal to α, the decision will be *reject H_O*.

When the calculated p-value is larger than α, the decision will be *fail to reject H_O*.

8.73 a. Fail to reject H_O b. Reject H_O

The p-value approach uses the calculated test statistic to find the area under the curve that contains the calculated test statistic and any values "beyond" it, in the direction of the alternative hypothesis. This probability (area under the curve), based on the position of the calculated test statistic, is compared to the level of significance (α) for the test and a decision is made.

Rules for calculating the p-value

1) If H_a contains <, then the p-value = $P(z < z^*)$.
2) If H_a contains >, then the p-value = $P(z > z^*)$.
3) If H_a contains $\neq$, then the p-value = $2P(z > |z^*|)$.

The p-value can then be calculated by using the z* value with Table 3, or it can be found directly using Table 5 (Appendix B, in JES2-p523), or it can be found using a computer and statistical software.

8.75 a. p-value = $P(z > 1.48)$ = 0.5000 - 0.4306 = <u>0.0694</u>

b. p-value = $P(z < -0.85)$ = 0.5000 - 0.3023 = <u>0.1977</u>

c. p-value = $2 \cdot P(z > 1.17)$ = 2(0.5000 - 0.3790) = <u>0.2420</u>

d. p-value = $P(z < -2.11)$ = 0.5000 - 0.4826 = <u>0.0174</u>

e. p-value = $2 \cdot P(z > 0.93)$ = 2(0.5000 - 0.3238) = <u>0.3524</u>

8.77 a. **P** = $P(z > z^*)$ = 0.0582

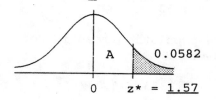

A   0.0582   A = 0.5000 - 0.0582 = 0.4418

0   z* = <u>1.57</u>

b. $P = P(z < z*) = 0.0166$

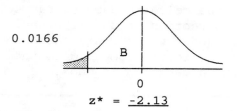

0.0166

$B = 0.5000 - 0.0166 = 0.4834$

$z* = \underline{-2.13}$

c. $P = P(z < -z*) + P(z > +z*) = 2 \cdot P(z > +z*) = 0.0042$

$$P(z > +z*) = 0.0021$$

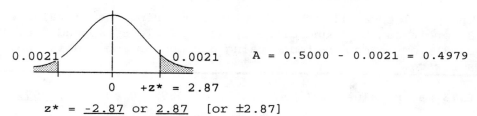

0.0021   0.0021 $A = 0.5000 - 0.0021 = 0.4979$

0 $+z* = 2.87$

$z* = \underline{-2.87}$ or $\underline{2.87}$ [or ±2.87]

MINITAB commands to complete a hypothesis test for a mean μ with a standard deviation σ can be found in JES2-p346.

Compare the calculated p-value to the given level of significance, α. Using the rules for comparison as stated in ST-p186, a decision can be made about the null hypothesis.

8.79   a. $H_O: \mu = 525$ vs. $H_a: \mu < 525$

b. Fail to reject H_O; the population mean is not significantly less than 525.

c. $\sigma_{\bar{x}} = \sigma/\sqrt{n} = 60.0/\sqrt{38} = 9.7333 = \underline{9.733}$

8.81 a. The mean test score for all elementary education majors.

b. H_O: $\mu = 35.70$
   H_a: $\mu < 35.70$

c. $z* = (32.63 - 35.70)/(6.73/\sqrt{165}) = -5.86$
   p-value $= P(z < -5.86) = P(z > 5.86) \approx +0.0000$

d. Reject H_O, the mean test score is less than 35.70 at the 0.001 level of significance.

8.83 Step 1: The average number of vaginal births per 100 women.
Step 2: H_O: $\mu = 12.6$
           H_a: $\mu > 12.6$
Step 3: a. normality assumed, CLT with n = 125
           b. z, $\sigma = 5.0$ c. $\alpha = 0.02$
Step 4: a. n = 125, $\overline{x}$ = 17.3
           b. z = $(\overline{x} - \mu)/(\sigma/\sqrt{n})$
               $z* = (17.3 - 12.6)/(5.0/\sqrt{125}) = 10.51$

-- 189 --

c. $P = P(z* > 10.51)$;
      Using Table 3, Appendix B, JES2-p521:
      $P = 0.5000 - 0.4999+ = 0.0000+$
      Using Table 5, Appendix B, JES2-p523:
      $P = 0.0000+$

Step 5: a. $P < \alpha$ b. Reject H_O
      c. At the 0.02 level of significance, there is sufficient evidence to support the contention that the mean number of vaginal births per 100 women is greater than 12.6.

SECTION 8.5 MARGIN EXERCISES

8.85 H_O: The mean shearing strength is at least 925 lbs.
     H_a: The mean shearing strength is less than 925 lbs.

8.86 H_O: $\mu = 9$ $(\leq)$
     H_a: $\mu > 9$

8.87 a. H_O: $\mu = 1.25$ $(\geq)$ b. H_O: $\mu = 335$
       H_a: $\mu < 1.25$ H_a: $\mu \neq 335$

     c. H_O: $\mu = 230,000$ $(\leq)$
       H_a: $\mu > 230,000$

8.88 Type A correct decision: The mean shearing strength is at least 925 lbs and it is decided that it is at least 925 lbs.
Type I error: The mean shearing strength is at least 925 lbs and it is decided that it is less than 925 lbs.
Type II error: The mean shearing strength is less than 925 lbs and it is decided that it is greater than or equal to 925 lbs.
Type B correct decision: The mean shearing strength is less than 925 lbs and it is decided that it is less than 925 lbs.

Type II error; you buy and use weak rivets.

8.89 $z = (\overline{x} - \mu)/(\sigma/\sqrt{n})$

$z^* = (354.3 - 356)/(17/\sqrt{120}) = \underline{-1.10}$

8.90 a. Reject H_O b. Fail to reject H_O

8.91

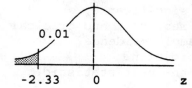

8.92

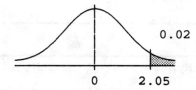

1 kg ≈ 2.2046 lbs

8.93 54.4 kg = 54.4(2.2046lbs) ≈ 119.9 or 120 lbs.

8.94 Results will vary, however expect your results to be similar to those shown in Table 8-10.

THE CLASSICAL HYPOTHESIS TEST: A FIVE-STEP PROCEDURE

Step 1: Describe the population parameter of concern.
Step 2: State the null hypothesis (H_O) and the alternative hypothesis (H_a).
Step 3: Specify the test criteria.
    a. Check the assumptions.
    b. Identify the test statistic to be used.
    c. Determine the level of significance, α.
    d. Determine the critical region(s) and critical value(s).
Step 4: Collect and present the sample evidence.
    a. Collect the sample information.
    b. Calculate the value of the test statistic.
Step 5: Determine the results.
    a. Determine whether or not the calculated test statistic is in the critical region.
    b. Make a decision about H_O.
    c. Write a conclusion about H_a.

Review "The Null and Alternative Hypotheses, H_O and H_a" in ST-p184, if necessary.

8.95 a. $H_O: \mu = 16$ yrs ($\geq$)   vs. $H_a: \mu < 16$

    b. $H_O: \mu = 6$ ft 6 in ($\leq$)   vs. $H_a: \mu > 6$ ft 6 in

    c. $H_O: \mu = 285$ ft ($\geq$)   vs. $H_a: \mu < 285$

    d. $H_O: \mu = 0.375$ inches ($\leq$)   vs. $H_a: \mu > 0.375$

    e. $H_O: \mu = 200$ units   vs. $H_a: \mu \neq 200$

> Critical region - that part under the curve where H_o will be
> rejected (size based on α)
> Noncritical region - the remaining part under the curve where H_o
> will not be rejected
> Critical value(s) - the $z(\alpha)$ or boundary point values of z,
> separating the critical and noncritical regions
>
> See JES2-p355 for a visual display of these regions and value(s).

8.97   a. The critical region is the set of all values of the test
statistic that will cause us to reject H_o.

b. The critical value(s) is the value(s) of the test
statistic that forms the boundary between the critical
region and the non-critical region. The critical value
is in the critical region.

> ### Determining the test criteria
>
> 1. Draw a picture of the standard normal (z) curve.
>   (0 is at the center)
>
> 2. Locate the critical region (based on α and H_a)
>   a) if H_a contains <, all of the α is placed in the left tail
>   b) if H_a contains >, all of the α is placed in the right tail
>   c) if H_a contains $\neq$, place $\alpha/2$ in each tail.
>
> 3. Shade in the critical region (the area where you will reject H_o).
>
> 4. Find the appropriate critical value(s) using the $z(\alpha)$ concept
>   and the Standard Normal Distribution (Table 3, Appendix B, JES2-
>   p521, or Table 4(a) for one-tail and Table 4(b) for two-tails,
>   Appendix B, JES2-p522).
>   If H_a contains >, the critical value is $z(\alpha)$.
>   If H_a contains <, the critical value is $z(1-\alpha)$ <u>or</u> $-z(\alpha)$.
>   If H_a contains $\neq$, the critical values are $\pm z(\alpha/2)$ <u>or</u> $z(\alpha/2)$
>   with $z(1-\alpha/2)$.
>
> Remember this boundary value divides the area under the
> curve into critical and noncritical regions and is part of
> the critical region.

8.99 a.

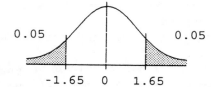

0.05   0.05

-1.65   0   1.65

b.

0.01

0   2.33

c.

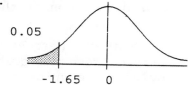

0.05

-1.65   0

d.

0.005   0.005

-2.58   0   2.58

For exercises 8.101 and 8.102, use $z* = (\bar{x} - \mu)/(\sigma/\sqrt{n})$, substituting the given values, then solving for the required unknown.

8.101 $z* = (\bar{x} - \mu)/(\sigma/\sqrt{n})$

$-1.18 = (\bar{x} - 250)/(22.6/\sqrt{85})$

$-1.18 = (\bar{x} - 250)/2.451314$

$-2.89255 = \bar{x} - 250$

$\bar{x} = 247.107449 = \underline{247.1}$

$\bar{x} = \Sigma x/n$

$247.107449 = \Sigma x/85$

$\Sigma x = \underline{21,004.133}$

NOTE: Standard Error $= \sigma_{\bar{x}} = \sigma/\sqrt{n}$ and $z = (\bar{x} - \mu)/(\sigma/\sqrt{n})$

8.103 a.   z = n(standard errors from mean):

$z = (\bar{x} - \mu)/(\sigma/\sqrt{n})$

$z = (4.8 - 4.5)/(1.0/\sqrt{100}) = \underline{3.0}$

$\underline{\bar{x} = 4.8 \text{ is } 3.0 \text{ standard errors } \textbf{above} \text{ the mean } \mu = 4.5}$

   b.   If $\alpha = 0.01$, the critical region is $z \geq 2.33$. Since z* is equal to 3.00, it is in the critical region. Therefore, yes, <u>reject H_o</u>.

RESULTS, DECISIONS AND CONCLUSIONS

Since the null hypothesis, H_O, is usually thought to be the statement whose truth is being challenged by the experimenter, all decisions are about the null hypothesis. The alternative hypothesis, H_a, however is usually thought to express the experimenter's viewpoint. Thus, the conclusion (interpretation of the decision) is expressed from the experimenter and alternative hypothesis point of view.

The two possible outcomes are:
1. z^* falls in the critical region or
2. z^* falls in the noncritical region.

Decision and Conclusion:

If z^* falls in the critical region, we **reject H_O**. The conclusion is very strong and proclaims the alternative to be the case, that is, there is sufficient evidence to overturn H_O in favor of H_a. It should read something like "There is sufficient evidence at the α level of significance to show that ...(the meaning of the H_a)."

If z^* falls in the noncritical (acceptance) region, we **fail to reject H_O**. The conclusion is much weaker, that is, it suggests that the data does not provide sufficient evidence to overturn H_O. This does not necessarily mean that we have to accept H_O at this point, but only that this sample did not provide sufficient evidence to reject H_O. It should read something like "There is not sufficient evidence at the α level of significance to show that ...(the meaning of the H_a)."

8.105 a. *Reject H_O* or *Fail to reject H_O*

      b. When the calculated test statistic falls in the critical region, the decision will be *reject H_O*.

        When the calculated test statistic falls in the non-critical region, the decision will be *fail to reject H_O*.

MINITAB commands to complete a hypothesis test using the classical approach can be found in JES2-p346. It is the same command used for the probability approach.
Compare the calculated z value (test statistic) with the corresponding critical value(s). The locations of z*, relative to the critical value of z, will determine the decision you must make about the null hypothesis.

8.107 a. H_O: $\mu = 15.0$ vs. H_a: $\mu \neq 15.0$

b. Critical values: $\pm z(0.005) = \pm 2.58$
Decision: reject H_O
Conclusion: There is sufficient evidence to conclude that the mean is different than 15.0.

c. $\sigma_{\bar{x}} = \sigma/\sqrt{n} = 0.5/\sqrt{30} = 0.091287 = \underline{0.0913}$

See ST-p189 for information on "Word Problems", if necessary.

<u>Hint for writing the hypotheses for exercise 8.109</u>
Look at the first sentence in the exercise, "The manager at Air Express feels ... are <u>less than</u> in the past." Since <u>less than</u> (<) does not include the equal to, it must be used in the alternative hypothesis. The negation is "...NOT less than...", which is > or =. Therefore, equality (=) is used in the null hypothesis (as usual), but it stands for greater than or equal to (≥).

8.109 Step 1: The mean weight of packages shipped by Air Express
Step 2: H_O: $\mu = 36.7$ (≥)
H_a: $\mu < 36.7$
Step 3: a. normality assumed, CLT with n = 64
b. z, $\sigma = 14.2$ c. $\alpha = 0.01$
d. $-z(0.01) = -2.33$

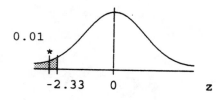

Step 4: a. n = 64, $\overline{x}$ = 32.1
   b. z = $(\overline{x} - \mu)/(\sigma/\sqrt{n})$
      z* = $(32.1 - 36.7)/(14.2/\sqrt{64})$ = -2.59
Step 5: a. z* falls in the critical region, see Step 3d
   b. Reject H_O
   c. At the 0.01 level of significance, the population
      mean is significantly less than the claimed mean
      of 36.7.

8.111 Step 1: The mean length of stay in days for non-government
         not-for-profit hospitals
Step 2: H_O: μ = 7.0 (≥)
        H_a: μ < 7.0
Step 3: a. normality assumed, CLT with n = 40
        b. z, σ = 1.5   c. α = 0.05
        d. -z(0.05) = -1.65

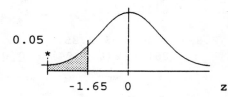

0.05

-1.65 0   z

Step 4: a. n = 40, $\overline{x}$ = 6.1
   b. z = $(\overline{x} - \mu)/(\sigma/\sqrt{n})$
      z* = $(6.1 - 7.0)/(1.5/\sqrt{40})$ = -3.79
Step 5: a. z* falls in the critical region, see Step 3d
   b. Reject H_O
   c. At the 0.05 level of significance, the sample
      provides sufficient evidence to conclude μ < 7.0
      days.

CHAPTER EXERCISES

8.113 a. $\overline{x}$ = <u>32.0</u>

b. σ = <u>2.4</u>

c. n = <u>64</u>

d. 1 - α = <u>0.90</u>

e. $z(\alpha/2) = z(0.05) = \underline{1.65}$

f. $\sigma_{\bar{x}} = 2.4/\sqrt{64} = \underline{0.3}$

g. $E = z(\alpha/2)\cdot\sigma/\sqrt{n} = (1.65)(0.3) = \underline{0.495}$

h. $UCL = \bar{x} + E = 32.0 + 0.495 = \underline{32.495}$

i. $LCL = \bar{x} - E = 32.0 - 0.495 = \underline{31.505}$

8.115 a. $H_O: \mu = 100$ b. $H_a: \mu \neq 100$

c. $\alpha = \underline{0.01}$ d. $\mu = \underline{100}$

e. $\bar{x} = \underline{96}$ f. $\sigma = \underline{12}$

g. $\sigma_{\bar{x}} = 12/\sqrt{50} = 1.697 = \underline{1.70}$

h. $z^* = (\bar{x} - \mu)/(\sigma/\sqrt{n}) = (96-100)/1.7 = \underline{-2.35}$

i. p-value $= 2\cdot P(z < -2.35) = 2\cdot P(z > 2.35)$
$= 2(0.5000 - 0.4906) = 2(0.0094) = \underline{0.0188}$

j. Fail to reject H_O

k. p-value $= 0.0188$; $\alpha = 0.01$

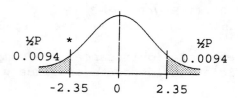

8.117 a. Step 1: The mean, μ
Step 2: a. normality assumed, CLT with n = 100
    b. z, $\sigma = 5.0$ c. $1-\alpha = 0.95$
Step 3: a. n = 100, $\bar{x} = 40.6$
    b. $\bar{x} = 40.6$

Step 4: a. $\alpha/2 = 0.05/2 = 0.025$; $z(0.025) = 1.96$
   b. $E = z(\alpha/2) \cdot \sigma/\sqrt{n} = (1.96)(5/\sqrt{100})$
      $= (1.96)(0.50) = 0.98$
   c. $\bar{x} \pm E = 40.6 \pm 0.98$
      <u>39.6 to 41.6</u>, the 0.95 confidence interval for μ

b. Step 1: The mean, μ
   Step 2: H_O: $\mu = 40$
      H_a: $\mu \neq 40$
   Step 3: a. normality assumed, CLT with n = 100
      b. z, $\sigma = 5.0$ c. $\alpha = 0.05$
   Step 4: a. n = 100, $\bar{x} = 40.6$
      b. $z = (\bar{x} - \mu)/(\sigma/\sqrt{n})$
         $z* = (40.6 - 40)/(5/\sqrt{100}) = 1.20$
      c. $P = 2P(z* > 1.20)$;
         Using Table 3, Appendix B, JES2-p521:
         $P = 2(0.5000 - 0.3849) = 2(0.1151) = 0.2302$
         Using Table 5, Appendix B, JES2-p523:
         $P = 2(0.1151) = 0.2302$
   Step 5: a. $P > \alpha$ b. Fail to reject H_O
      c. At the 0.05 level of significance, there is
         not sufficient evidence to support the
         contention that the mean is not equal to 40.

c. Step 1: The mean, μ
   Step 2: H_O: $\mu = 40$
      H_a: $\mu \neq 40$
   Step 3: a. normality assumed
      b. z, $\sigma = 5.0$ c. $\alpha = 0.05$
      d. $\pm z(0.025) = \pm 1.96$

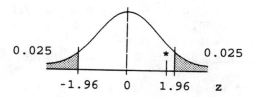

Step 4: a. n = 100, $\bar{x} = 40.6$
   b. $z = (\bar{x} - \mu)/(\sigma/\sqrt{n})$
      $z* = (40.6 - 40)/(5/\sqrt{100}) = 1.20$

Step 5: a. z* falls in the noncritical region, see * in
Step 3d.
b. Fail to reject H_O
c. At the 0.05 level of significance, there is
not sufficient evidence to support the
contention that the mean is not equal to 40.

d.

α = 0.05

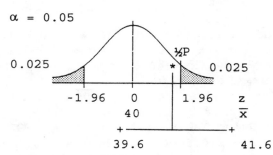

z^* = 1.20 is in the noncritical region or P = 0.2302 is
greater than α, and μ = 40 is within the interval estimate
of 39.6 to 41.6.

Exercise 8.119 shows the effect of the level of confidence (1 - α)
on the width of a confidence interval.

8.119 a. Step 1: The mean weights of full boxes of a certain kind
of cereal
Step 2: a. normality indicated
b. z, σ = 0.27   c. 1-α = 0.95
Step 3: a. n = 18, $\overline{x}$ = 9.87
b. $\overline{x}$ = 9.87
Step 4: a. α/2 = 0.05/2 = 0.025; z(0.025) = 1.96
b. E = z(α/2)·σ/$\sqrt{n}$ = (1.96)(0.27/$\sqrt{18}$)
= (1.96)(0.0636) = 0.12
c. $\overline{x}$ ± E = 9.87 ± 0.12
9.75 to 9.99, the 0.95 confidence interval for μ

b. Step 1: The mean weights of full boxes of a certain kind
of cereal
Step 2: a. normality indicated
b. z, σ = 0.27   c. 1-α = 0.99
Step 3: a. n = 18, $\overline{x}$ = 9.87
b. $\overline{x}$ = 9.87

Step 4: a. $\alpha/2 = 0.01/2 = 0.005$; $z(0.005) = 2.58$
   b. $E = z(\alpha/2)\cdot\sigma/\sqrt{n} = (2.58)(0.27/\sqrt{18})$
    $= (2.58)(0.0636) = 0.16$
   c. $\overline{x} \pm E = 9.87 \pm 0.16$
   <u>9.71 to 10.03</u>, the 0.99 confidence interval for μ

 c. The increased confidence level widened the interval.

8.121 a. Step 1: The mean score for a clerk-typist position
   Step 2: a. normality assumed, CLT with $n = 100$
     b. z, $\sigma = 10.5$ c. $1-\alpha = 0.99$
   Step 3: a. $n = 100$, $\overline{x} = 72.6$
     b. $\overline{x} = 72.6$
   Step 4: a. $\alpha/2 = 0.01/2 = 0.005$; $z(0.005) = 2.58$
     b. $E = z(\alpha/2)\cdot\sigma/\sqrt{n} = (2.58)(10.5/\sqrt{100})$
      $= (2.58)(1.05) = 2.71$
     c. $\overline{x} \pm E = 72.6 \pm 2.71$
     <u>69.89 to 75.31</u>, the 0.99 confidence interval for μ

 b. <u>Yes.</u> 75.0 falls within the interval.

8.123 Step 1: The mean efficacy expectation score for preoperative
   patients
  Step 2: a. normality assumed, CLT with $n = 200$
    b. z, $\sigma = 0.94$ c. $1-\alpha = 0.95$
  Step 3: a. $n = 200$, $\overline{x} = 4.00$
    b. $\overline{x} = 4.00$
  Step 4: a. $\alpha/2 = 0.05/2 = 0.025$; $z(0.025) = 1.96$
    b. $E = z(\alpha/2)\cdot\sigma/\sqrt{n} = (1.96)(0.94/\sqrt{200})$
     $= (1.96)(0.066) = 0.13$
    c. $\overline{x} \pm E = 4.00 \pm 0.13$
    <u>3.87 to 4.13</u>, the 0.95 confidence interval for μ

8.125 $n = [z(\alpha/2)\cdot\sigma/E]^2 = [(2.58)(\sigma)/(\sigma/3)]^2 = 59.9 = \underline{60}$

8.127 $z = (\overline{x} - \mu)/(\sigma/\sqrt{n})$

 $z* = (12.5 - 10.0)/(7.5/\sqrt{75}) = \underline{2.89}$

 P $= P(z > 2.89) = 0.5000 - 0.4981 = \underline{0.0019}$

8.129 a. H_O: $\mu = 300$ vs. H_a: $\mu < 300$

b. $z = (\overline{x} - \mu)/(\sigma/\sqrt{n})$

  $z^* = (295 - 300)/(24/\sqrt{45}) = \underline{-1.40}$

  $P = P(z < -1.40) = 0.5000 - 0.4192 = \underline{0.0808}$

c. $\alpha = 0.01$

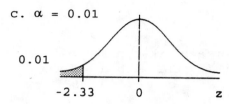

0.01

  -2.33   0   z

  $z = (\overline{x} - \mu)/(\sigma/\sqrt{n})$

  $-2.33 = (\overline{x} - 300)/(24/\sqrt{45})$

  $\overline{x} = 300 + (-2.33)(24/\sqrt{45}) = \underline{291.664}$

NOTE: In exercises 8.131 and 8.133, both methods of hypothesis testing are asked for. Several steps are the same for both methods. The indented section after Step 4 includes the specific parts for each method.

8.131 Step 1: The mean customer checkout time at a large
          supermarket
  Step 2: H_O: $\mu = 9$ ($\leq$) (no more than)
          H_a: $\mu > 9$
  Step 3: a. normality indicated
          b. z, $\sigma = 2.5$ c. $\alpha = 0.02$
  Step 4: a. n = 24, $\overline{x} = 10.6$
          b. $z = (\overline{x} - \mu)/(\sigma/\sqrt{n})$
             $z^* = (10.6 - 9.0)/(2.5/\sqrt{24}) = 3.14$

- -

  a.   Step 4c. $P = P(z^* > 3.14)$;
               Using Table 3, Appendix B, JES2-p521:
               $P = 0.5000 - 0.4992 = 0.0008$
               Using Table 5, Appendix B, JES2-p523:
               $0.0008 < P < 0.0010$
       Step 5a. $P < \alpha$
  OR

b. Step 3d. z(0.02) = 2.05

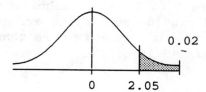

0.02

0   2.05

        Step 5a. z* falls in the critical region, see *
                 Step 3d

Step 5: b. Reject H$_O$
        c. At the 0.02 level of significance, the sample
           does provide sufficient evidence to conclude
           the mean waiting time is more than the claimed
           9 minutes.

8.133 Step 1: The mean diameter of rivets
        Step 2: H$_O$: $\mu = 0.3125$
                H$_a$: $\mu > 0.3125$
        Step 3: a. normality assumed, CLT with n = 280
                b. z, $\sigma = 0.0006$ c. $\alpha = 0.01$
        Step 4: a. n = 280, $\bar{x} = 0.3126$
                b. z = $(\bar{x} - \mu)/(\sigma/\sqrt{n})$
                   z* = $(0.3126 - 0.3125)/(0.0006/\sqrt{280})$ = 2.79
----
    a. Step 4c. **P** = P(z > 2.79);
                 Using Table 3, Appendix B, JES2-p521:
                 P = 0.5000 - 0.4974 = 0.0026
                 Using Table 5, Appendix B, JES2-p523:
                 0.0026 < **P** < 0.0030
        Step 5a. **P** < α
    OR
    b. Step 3d. z(0.01) = 2.33

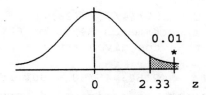

0.01
*

0   2.33   z

        Step 5a. z* falls in the critical region, see *
                 Step 3d

-- 203 --

Step 5: b. Reject H_O

    c. At the 0.01 level of significance, the sample does provide sufficient evidence to conclude the mean diameter is more than the claimed 0.3125 (5/16) inch.

8.135 a. 125.10 to 132.95, the 0.95 confidence interval for μ

  b. $\sigma_{\bar{x}} = \sigma / \sqrt{n} = 10.0 / \sqrt{25} = 2.00$

    $\bar{x} \pm z(\alpha/2) \cdot \sigma/\sqrt{n} = 129.02 \pm (1.96)(2.00)$
                      129.02 ± 3.92

    125.10 to 132.94, the 0.95 confidence interval for μ

8.137 a. H_a: $\mu \neq 18$; Fail to reject H_O; The population mean is not significantly different from 18.

  b. $\sigma_{\bar{x}} = \sigma / \sqrt{n} = 4.00 / \sqrt{28} = 0.756$

    $z = (\bar{x} - \mu)/(\sigma/\sqrt{n})$

    $z\star = (17.217 - 18)/(0.756) = -1.04$

    p-value $= 2 \cdot P(z\star < -1.04) = 2 \cdot P(z\star > 1.04)$
              $= 2(0.5000 - 0.3508) = 2(0.1492) = 0.2984 = \underline{0.30}$

8.139 Every student will have different results, but they should be similar to the following.
a. Commands needed:
     RANDom 50 C1-C28;
     NORMal 19 4.
     RMEAn C1-C28 C29
     LET C30=((C29-18)/(4/SQRT(28)))
b. In one run of the commands 34/50 or 68% of the values were more extreme than the given z-values; an empirical p-value.
c. critical points $= \pm 2.58$; 45/50 = 90% fell in the noncritical region; an empirical β, probability of type II error.

CHAPTER 9 ∇ INFERENCES INVOLVING ONE POPULATION

Chapter Preview

Chapter 9 continues the work of inferential statistics started in Chapter 8. The concepts of hypothesis tests and confidence intervals will still be presented but on samples where the population standard deviation (σ) is unknown. Also inferences regarding the population binomial probability (p) will be introduced.

SECTION 9.1 MARGIN EXERCISES

9.1 Pick any 3 numbers; the fourth must be the negative of the sum of the first three. For example; 4, 3, 1, whose sum is 8; the fourth is required to be -8 for the sum to be zero.

9.2 The bottom row of Table 6 is identical to the $z(\alpha)$ values in Table 4A. If σ is unknown, Table 6 is used, but note that for df > 100, the critical values for Student-t are approximately the same as those of the standard normal.

9.3 a. t(12, 0.01) = <u>2.68</u>   b. t(22, 0.025) = <u>2.07</u>

9.4 a. t(18, 0.90) = -t(18, 0.10) = <u>-1.33</u>
     b. t(9, 0.99) = -t(9, 0.01) = <u>-2.82</u>

9.5 $\alpha/2$ = 0.05/2 = 0.025;   ±t(12, 0.025) = <u>±2.18</u>

9.6 Step 1: The mean, μ
     Step 2: a. normality assumed
             b. t   c. $1-\alpha$ = 0.95
     Step 3: a. n = 24, $\overline{x}$ = 16.7, s = 2.6
             b. $\overline{x}$ = 16.7
     Step 4: a. $\alpha/2$ = 0.05/2 = 0.025; df = 23;
                 t(23, 0.025) = 2.07

b. $E = t(df, \alpha/2) \cdot (s/\sqrt{n}) = (2.07)(2.6/\sqrt{24})$
    $= 1.0986 = 1.10$
c. $\bar{x} \pm E = 16.7 \pm 1.10$
   <u>15.60 to 17.80</u>, the 0.95 confidence interval for μ

9.7 Step 1: The mean, μ
      Step 2: H_O: $\mu = 32$ ($\leq$)
              H_a: $\mu > 32$
      Step 3: a. normality assumed
              b. t   c. $\alpha = 0.05$
      Step 4: a. $n = 16$, $\bar{x} = 32.93$, $s = 3.1$
              b. $t = (\bar{x} - \mu)/(s/\sqrt{n})$
                 $t^* = (32.93 - 32)/(3.1/\sqrt{16}) = 1.20$
              c. $P = P(t > 1.20 | df = 15)$;
                 Using Table 6, Appendix B, JES2-p524:
                 $0.10 < P < 0.25$
                 Using Table 7, Appendix B, JES2-p525:
                 $P = 0.124$
      Step 5: a. $P > \alpha$ b. Fail to reject H_O
              c. At the 0.05 level of significance, there is
                 sufficient evidence that the mean is not greater
                 than 32.

9.8 Step 1: The mean, μ
      Step 2: H_O: $\mu = 73$
              H_a: $\mu \neq 73$
      Step 3: a. normality assumed
              b. t   c. $\alpha = 0.05$
      Step 4: a. $n = 12$, $\bar{x} = 71.46$, $s = 4.1$
              b. $t = (\bar{x} - \mu)/(s/\sqrt{n})$
                 $t^* = (71.46 - 73)/(4.1/\sqrt{12}) = -1.30$
              c. $P = P(t < -1.30 | df = 11) + P(t > 1.30 | df = 11)$
                 $= 2P(t > 1.30 | df = 11)$;
                 Using Table 6, Appendix B, JES2-p524:
                 $0.20 < P < 0.50$
                 Using Table 7, Appendix B, JES2-p525:
                 $P = 2(0.110) = 0.220$
      Step 5: a. $P > \alpha$ b. Fail to reject H_O
              c. At the 0.05 level of significance, there is
                 sufficient evidence that the mean is not
                 significantly different from 73.

9.9 Using Table 6: $P = P(t > 1.08 | df = 16)$; $0.10 < P < 0.25$
Using Table 7: $P = P(t > 1.08 | df = 16)$; $0.143 < P < 0.167$

9.10 Step 1: The mean, μ
Step 2: H_O: $\mu = 52$ $(\geq)$
        H_a: $\mu < 52$
Step 3: a. normality assumed
        b. t   c. $\alpha = 0.01$
        d. $-t(9, 0.01) = -2.82$

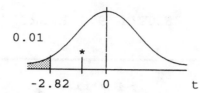

0.01

-2.82   0   t

Step 4: a. $n = 10$, $\overline{x} = 50.47$, $s = 3.9$
        b. $t = (\overline{x} - \mu)/(s/\sqrt{n})$
           $t* = (50.47 - 52)/(3.9/\sqrt{10}) = -1.24$
Step 5: a. $t*$ falls in the noncritical region, see Step 3d
        b. Fail to reject H_O
        c. At the 0.01 level of significance, there is not
           sufficient evidence that the mean is less than 52.

9.11 $\alpha = 0.02$
$\alpha/2 = 0.01$
$\pm t(54, 0.01) = \pm 2.40$

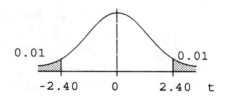

0.01   0.01

-2.40   0   2.40   t

9.12 a. $P = P(t > 1.92 | df = 44)$
        using Table 6: $0.025 < P < 0.05$
        using Table 7: $0.026 < P < 0.033$
        $\alpha = 0.05$, $P \leq \alpha$, therefore it is significant

        b. $P = P(t > 3.41 | df = 44)$
        using Table 6: $P < 0.005$ [not enough information given]
        using Table 7: $P = 0.001$
        $\alpha = 0.001$, $P \leq \alpha$, therefore it is significant

c. $P = P(t > 1.81 | df = 44)$ - two-tailed
   using Table 6: $0.050 < P < 0.100$
   using Table 7: $0.064 < P < 0.080$

To be significant, the smallest possible p-value is desired. For a two-tailed test and Table 6, $0.05 < p < 0.10$, i.e., $p < 0.10$ would be reported. For a one-tailed test, $0.025 < p < 0.05$, i.e., $p < 0.05$ would be reported.

SECTION 9.1 EXERCISES

---

t-Distribution
(used when σ is unknown)

Key facts about the t-distribution:

1. The total area under the t-distribution is 1.

2. It is symmetric about 0.

3. Its shape is a more "spread out" version of the normal shape.

4. A different curve exists for each sample size.

5. The shape of the distribution approaches the normal distribution shape as n increases [For df > 100, t is approximately normal.].

6. Critical values are determined based on α and degrees of freedom(df) - Table 6 (Appendix B, JES2-p524).

7. Degrees of freedom is abbreviated as *df*, where df = n - 1 for this application.

Notation: t(df,α) = t(degrees of freedom, area to the right)

            ↑   ↑   ↑

            Table 6   row id #   column id #

ex.: t(13,.025) means df = 13 (row) and α = .025 (column), using

        Table 6, t(13,.025) = 2.16   (df = n-1)

For t(df,α), consider the α given as the amount in one tail and use the top row label - "Amount of α in One-Tail". For two-tailed tests, an additional row label is given - "Amount of α in Two-Tails". Note that it is twice the amounts in the one-tail row, therefore α does not have to be divided by two.

For α > 0.5000, use the 1-α amount and negate the t-value.

ex.: t(14,0.90); α = 0.90, 1-α = 0.10,

               t(14,0.90) = -t(14,0.10) = -1.35

(Table 6 is in Appendix B, JES2-p524.)

9.13 a. 1.71   b. 1.37   c. 2.60   d. 2.08

       e. -1.72   f. -2.06   g. -2.47   h. 2.01

For a two-sided test:

    1. divide α by 2 and use the top row of column labels of Table 6 (Appendix B, JES2-p524) identified as "Amount of α (α/2, in this case) in One-Tail"

    or

    2. use the second row of column labels of Table 6 identified as "Amount of α in Two-Tails."

Drawing a diagram of a t-curve and labeling the regions with the given information will be helpful in answering exercises 9.15 through 9.18.

9.15 df = 7

9.17 a. -2.49   b. 1.71   c. -0.685

9.19 a. Symmetric about mean: mean is 0

  b. Standard deviation of t-distribution is greater than 1;
  t-distribution is different for each different sample
  size-while there is only one z-distribution.

Estimating μ - the population mean

(σ unknown)

1. point estimate: $\bar{x}$

2. confidence interval: $\bar{x} \pm t(df,\alpha/2) \cdot (s/\sqrt{n})$, where df = n-1

Review steps for constructing a confidence interval for μ: JES2-
p315, ST-p175. The t-distribution is used when σ is unknown, and
sampling is from an approximately normal distribution or the sample
size is large.

9.21 Step 1: The mean age of onset of offending for those
         convicted of antitrust offenses
  Step 2: a. Age distribution presumed normal for this offense
          b. t   c. 1-α = 0.90
  Step 3: a. n = 35, $\bar{x}$ = 54, s = 7.5
          b. $\bar{x}$ = 54
  Step 4: a. α/2 = 0.10/2 = 0.05; df = 34; t(34, 0.05) = 1.70
          b. E = t(df,α/2)·(s/$\sqrt{n}$) = (1.70)(7.5/$\sqrt{35}$)
                = 2.15514 = 2.16
          c. $\bar{x}$ ± E = 54 ± 2.16
             <u>51.84 to 56.16</u>, the 0.90 confidence interval for μ

9.23 Step 1: The mean gestation period for mothers who meet the 7
         eligibility requirements
  Step 2: a. The sampled population is assumed to be normally
             distributed.
          b. t   c. 1-α = 0.95
  Step 3: a. n = 20, $\bar{x}$ = 40, s = 1.5
          b. $\bar{x}$ = 40
  Step 4: a. α/2 = 0.05/2 = 0.025; df = 19;
             t(19, 0.025) = 2.09

b. $E = t(df,\alpha/2) \cdot (s/\sqrt{n}) = (2.09)(1.5/\sqrt{20})$
    $= 0.701 = 0.7$
c. $\overline{x} \pm E = 40 \pm 0.7$
    <u>39.3 to 40.7</u>, the 0.95 confidence interval for μ

9.25 a. $\overline{x} = \Sigma x/n = 550.22/41 = \underline{13.42}$

b. $s = \sqrt{[\Sigma(x - \overline{x})^2 / (n - 1)]} = \sqrt{1617.984/40} = \underline{6.36}$

c. Step 1: The mean textbook cost per semester
   Step 2: a. normality assumed, CLT with n = 41
           b. t   c. $1 - \alpha = 0.90$
   Step 3: a. $n = 41$, $\overline{x} = 13.42$, $s = 6.36$
           b. $\overline{x} = 13.42$
   Step 4: a. $\alpha/2 = 0.10/2 = 0.05$; $df = 40$;
              $t(40, 0.05) = 1.68$
           b. $E = t(df,\alpha/2) \cdot (s/\sqrt{n}) = (1.68)(6.36/\sqrt{41}) = 1.67$
           c. $\overline{x} \pm E = 13.42 \pm 1.67$
              <u>11.75 to 15.09</u>, the 0.90 confidence interval for μ

MINITAB commands to calculate a confidence interval for the
population mean, μ, when the population standard deviation, σ, is
unknown can be found in JES2-p381.

9.27 MINITAB verify - answers given in exercise.

Hypotheses are written the same way as before. Sample size and
standard deviation have no effect on the stating of hypotheses.

9.29 a. $H_O: \mu = 11$ ($\geq$) vs. $H_a: \mu < 11$

b. $H_O: \mu = 54$ ($\leq$) vs. $H_a: \mu > 54$

c. $H_O: \mu = 75$ vs. $H_a: \mu \neq 75$

9.31 a. α = 0.05

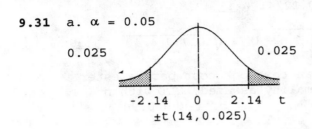

b. α = 0.01

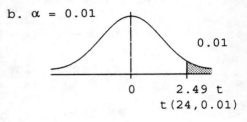

c. α = 0.05

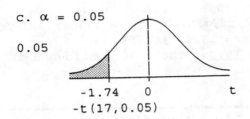

d. α = 0.01

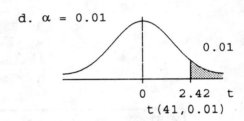

Calculating the **P**-value using the t-distribution

Table 6 or Table 7 (Appendix B, JES2-pp524&525) or a computer, can be used to <u>estimate</u> the p-value

1. Using Table 6 to place bounds on the value of **P**
   - a) locate df row
   - b) locate the absolute value of the calculated t-value between two critical values in the df row
   - c) the p-value is in the interval between the two corresponding probabilities at the top of the columns; read the bounds from either the *one-tail* or *two-tailed* column headings as per H_a.

2. Using Table 7 to estimate or place bounds on the value of **P**
   - a) locate the absolute value of the calculated t-value and the df directly for the corresponding probability value

   OR

   - b) locate the absolute value of the calculated t-value and its df between appropriate bounds. From the box formed, use the upper left and lower right values for the interval.
     (see JES2-p385)

3. Using a computer
   - a) the p-value is calculated directly and given in the output when using the TTESt command on MINITAB

   OR

   - b) the p-value is calculated using the CDF command:
        CDF (t-value);
        T with (n-1) df.

     Subtract the probability value from 1 or multiply it by 2, depending on the exercise. The cumulative probability given is $P(t \leq$ t-value$)$.

-- 213 --

9.33 a. P-value approach:

Step 1: The mean travel time to college

Step 2: H_o: $\mu = 25$ (at least) (≥)

   H_a: $\mu < 25$ (less than)

Step 3: a. normality assumed, CLT with n = 22

   b. t   c. $\alpha = 0.01$

Step 4: a. n = 22, $\overline{x} = 19.4$,   s = 9.6

   b. $t = (\overline{x} - \mu)/(s/\sqrt{n})$

   $t* = (19.4 - 25.0)/(9.6/\sqrt{22}) = -2.74$

   c. $P = P(t < -2.74 | df = 21) = P(t > 2.74 | df = 21)$;

   Using Table 6, Appendix B, JES2-p524:

   $0.005 < P < 0.01$

   Using Table 7, Appendix B, JES2-p525:

   $0.005 < P < 0.007$

Step 5: a. $P < \alpha$ b. Reject H_o

   c. At the 0.01 level of significance, the sample
   does provide sufficient evidence to justify the
   contention that mean travel time is less than 25
   minutes.

┌───┐
│   │
│ Hypothesis tests (classical approach) will be completed using the │
│ same format as before. You may want to review: JES2-p350, ST-p192. │
│ The only differences are in: │
│   1. finding the critical value(s) of t   │
│   remember you need α (column) and degrees of │
│   freedom (df = n - 1) (row) for the t-distribution │
│   │
│   2. the calculated test statistic, which is t, where   │
│ $t = (\overline{x} - \mu)/(s/\sqrt{n})$.   │
│   │
└───┘

b. Classical approach:

Step 1: The mean travel time to college
Step 2: H_O: $\mu = 25$ (at least) ($\geq$)
$\quad\quad\quad$ H_a: $\mu < 25$ (less than
Step 3: a. Travel times are mounded; assume normality,
$\quad\quad\quad\quad$ CLT with n = 22
$\quad\quad\quad$ b. t $\quad\quad\quad\quad$ c. $\alpha = 0.01$
$\quad\quad\quad$ d. $-t(21, 0.01) = -2.52$

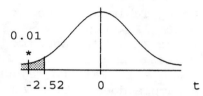

$\quad\quad$ 0.01

$\quad\quad\quad$ -2.52 $\quad\quad$ 0 $\quad\quad\quad\quad$ t

Step 4: a. n = 22, $\bar{x}$ = 19.4, $\quad$ s = 9.6
$\quad\quad\quad$ b. $t = (\bar{x} - \mu)/(s/\sqrt{n})$
$\quad\quad\quad\quad$ $t* = (19.4 - 25.0)/(9.6/\sqrt{22}) = -2.74$
Step 5: a. t* falls in the critical region, see * Step 3d
$\quad\quad\quad$ b. Reject H_O
$\quad\quad\quad$ c. At the 0.01 level of significance, the sample
$\quad\quad\quad\quad$ does provide sufficient evidence to justify the
$\quad\quad\quad\quad$ contention that mean travel time is less than 25
$\quad\quad\quad\quad$ minutes.

$\quad\quad\quad$ **Hint for writing the hypotheses for exercise 9.35**
Look at the last sentence of the exercise, "Assuming that the
sample .. test the hypothesis of '<u>different from</u>' using a level of
significance of 0.05". The words "different from" indicate a
different than or <u>not equal to</u> ($\neq$). The negation becomes "equal
to" and the null hypothesis would be written with an equality sign
($=$).

9.35 a. The "population" data ranged from 6% to 71.6%, therefore
$\quad\quad\quad$ the midrange is 38.8%. When the midrange is close in
$\quad\quad\quad$ value to the mean, the distribution is approximately
$\quad\quad\quad$ symmetrical; therefore, the assumption of normality is
$\quad\quad\quad$ reasonable.

b. P-value approach:

Step 1: The mean percentage intake of kilocalories from fat
Step 2: $H_{O:}$ $\mu = 38.4\%$
$\quad\quad$ H_a: $\mu \neq 38.4\%$
Step 3: a. normality indicated
$\quad\quad$ b. t $\quad\quad\quad$ c. $\alpha = 0.05$
Step 4: a. $n = 15$, $\bar{x} = 40.5\%$, $\quad$ $s = 7.5\%$
$\quad\quad$ b. $t = (\bar{x} - \mu)/(s/\sqrt{n})$
$\quad\quad\quad$ $t* = (40.5 - 38.4)/(7.5/\sqrt{15}) = 1.08$
$\quad\quad$ c. $P = 2P(t > 1.08 | df = 14)$;

$\quad\quad\quad$ Using Table 6, Appendix B, JES2-p524:
$\quad\quad\quad$ $0.20 < P < 0.50$
$\quad\quad\quad$ Using Table 7, Appendix B, JES2-p525:
$\quad\quad\quad$ $0.144 < \frac{1}{2}P < 0.169$; $\quad$ $0.288 < P < 0.338$
Step 5: a. $P > \alpha$ $\quad\quad$ b. Fail to reject H_O
$\quad\quad$ c. The sample does not provide sufficient evidence
$\quad\quad\quad$ to justify the contention that the mean
$\quad\quad\quad$ percentage is different than 38.4%, at the 0.05
$\quad\quad\quad$ level of significance.

c. Classical approach:

Step 1: The mean percentage intake of kilocalories from fat
Step 2: $H_{O:}$ $\mu = 38.4\%$
$\quad\quad$ H_a: $\mu \neq 38.4\%$
Step 3: a. normality indicated
$\quad\quad$ b. t $\quad\quad\quad\quad$ c. $\alpha = 0.05$
$\quad\quad$ d. $\pm t(14, 0.025) = \pm 2.14$

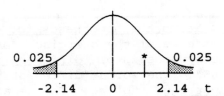

$\quad\quad$ 0.025 $\quad\quad\quad\quad$ * $\quad$ 0.025
$\quad\quad$ -2.14 $\quad\quad$ 0 $\quad\quad$ 2.14 $\quad$ t

Step 4: a. $n = 15$, $\bar{x} = 40.5\%$, $\quad$ $s = 7.5\%$
$\quad\quad$ b. $t = (\bar{x} - \mu)/(s/\sqrt{n})$
$\quad\quad\quad$ $t* = (40.5 - 38.4)/(7.5/\sqrt{15}) = 1.08$
Step 5: a. $t*$ falls in the noncritical region, see * Step 3d
$\quad\quad$ b. Fail to reject H_O
$\quad\quad$ c. The sample does not provide sufficient evidence
$\quad\quad\quad$ to justify the contention that the mean
$\quad\quad\quad$ percentage is different than 38.4%, at the 0.05
$\quad\quad\quad$ level of significance.

9.37 Sample statistics: $n = 6$, $\Sigma x = 222$, $\Sigma x^2 = 8330$,
$$\overline{x} = 37.0, \quad s = 4.817$$

     a. P-value approach:

        Step 1: The mean test score at a certain university
        Step 2: H_0: $\mu = 35$ (reasonable)
              H_a: $\mu \neq 35$ (not reasonable)
        Step 3: a. normality indicated
             b. t   c. $\alpha = 0.05$
        Step 4: a. $n = 6$, $\overline{x} = 37.0$, $s = 4.817$
             b. $t = (\overline{x} - \mu)/(s/\sqrt{n})$
               $t* = (37.0 - 35.0)/(4.817/\sqrt{6}) = 1.02$
             c. $P = 2P(t > 1.02 | df = 5)$;
               Using Table 6, Appendix B, JES2-p524:
               $0.20 < P < 0.50$
               Using Table 7, Appendix B, JES2-p525:
               $0.161 < \tfrac{1}{2}P < 0.182$] $0.322 < P < 0.364$
        Step 5: a. $P > \alpha$ b. Fail to reject H_0
             c. The sample does not provide sufficient
               evidence to reject the claim that the mean
               score is 35, at the 0.05 level of
               significance.

    b. Classical approach:

      Step 1: The mean test score at a certain university
      Step 2: H_0: $\mu = 35$ (reasonable)
           H_a: $\mu \neq 35$ (not reasonable)
      Step 3: a. normality indicated
            b. t   c. $\alpha = 0.05$
            d. $\pm t(5, 0.025) = \pm 2.57$

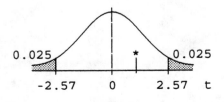

      0.025   *   0.025
       -2.57   0   2.57 t

Step 4: a. n = 6, $\bar{x}$ = 37.0, s = 4.817

      b. $t = (\bar{x} - \mu)/(s/\sqrt{n})$

        $t* = (37.0 - 35.0)/(4.817/\sqrt{6}) = 1.02$

Step 5: a. t* falls in the noncritical region, see * Step 3d

      b. Fail to reject H_O

      c. The sample does not provide sufficient evidence
to reject the claim that the mean score is 35, at
the 0.05 level of significance.

MINITAB command to perform a hypothesis test for a population mean
if the population standard deviation (σ) is unknown can be found in
JES2-p385.

The alternative command works the same as in the ZTEST command.
The output will also look the same except a t-value will be
calculated in place of the z-value and the SEMEAN is an estimated
standard error, since it is using $s/\sqrt{n}$.

9.39 MINITAB verify - answers given in exercise.

9.41 Results will vary.

    Suggestion: use these additional session subcommands
following the subcommand CUTPoint each time you construct a
histogram: BAR; SYMBOL; TYPE 0; LABEL.

    For the normal population:

    A histogram of 1000 random data presents a good picture of
the population. Notice, the data ranges from -75 to 250.

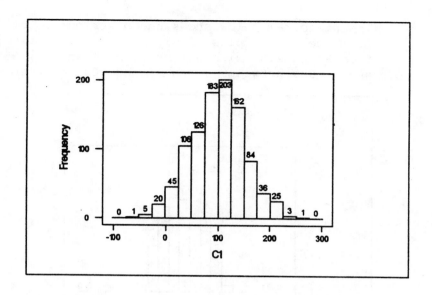

The histogram below shows the 100 sample means from samples
of n = 10 from N(100,50). Notice, the x-bars range from 60
to 130, a much smaller range than the x-values above. This
histogram shows 72% of the x-bars are within approximately
one standard error of the mean; 95% within two and 100%
within three.

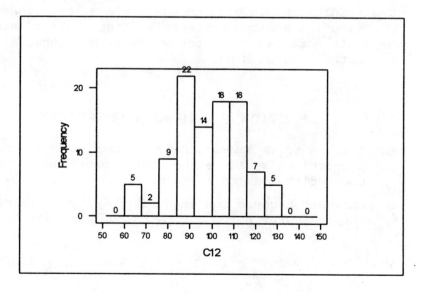

The histogram below shows the distribution of 100 t* values
from the 100 samples of size 10 above. Notice, t* ranges
from -4 to 2.5, with 65% between -1 and +1, 95% between -2
and 2, 98% between -3 and 3, and 100% between -4 and 4.

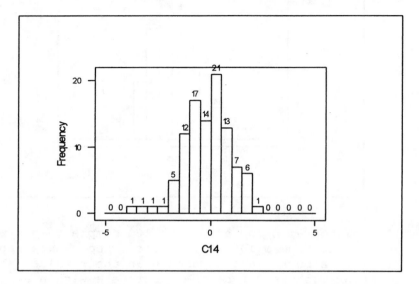

Your results will vary but will tend to be similar.

The results obtained using the uniform (or rectangular)
distribution and the exponential distribution (skewed) will
be a bit more variable, but you will be amazed at the
predictability of x-bar.

SECTION 9.2 MARGIN EXERCISES

9.42 a. Yes, it seems likely that the mean of the observed
proportions would be the true proportion for the
population.(law of large numbers)

b. Unbiased because the mean of the p' distribution is p,
the parameter being estimated.

9.43 $\sqrt{npq}/n = \sqrt{npq}/\sqrt{n^2} = \sqrt{npq/n^2} = \sqrt{pq/n}$

9.44 Step 1: The proportion of convertibles driven by students
Step 2: a. The sample was randomly selected and each subject's response was independent of those of the others surveyed.

b. $n = 400$; $n > 20$, $np = (400)(92/400) = 92$, $nq = (400)(308/400) = 308$, np and nq both > 5

c. $1 - \alpha = 0.95$

Step 3: a. $n = 400$, $x = 92$

b. $p' = x/n = 92/400 = 0.23$

Step 4: a. $z(\alpha/2) = z(0.025) = 1.96$

b. $E = z(\alpha/2) \cdot \sqrt{p'q'/n} = 1.96\sqrt{(0.23)(0.77)/400}$
$= (1.96)(0.02104) = 0.041$

c. $p' \pm E = 0.23 \pm 0.041$
      <u>0.189 to 0.271</u> is the 0.95 interval for
                  $p = P(\text{drives convertible})$

9.45 Step 1: $1 - \alpha = 0.95$; $z(\alpha/2) = z(0.025) = 1.96$
Step 2: $E = 0.02$
Step 3: no estimate given, $p* = 0.5$ and $q* = 0.5$
Step 4: $n = \{[z(\alpha/2)]^2 \cdot p* \cdot q*\}/E^2$
$= (1.96^2)(0.5)(0.5)/(0.02^2) = \underline{2401}$

9.46 Step 1: $1 - \alpha = 0.90$; $z(\alpha/2) = z(0.05) = 1.65$
Step 2: $E = 0.02$
Step 3: $p* = 0.25$ and $q* = 0.75$
Step 4: $n = \{[z(\alpha/2)]^2 \cdot p* \cdot q*\}/E^2$
$= (1.65^2)(0.25)(0.75)/(0.02^2) = 1276.17 = \underline{1277}$

9.47 Step 1: The proportion, p
Step 2: H_O: $p = 0.50$ $(\geq)$
        H_a: $p < 0.50$
Step 3: a. independence assumed

b. z; $n = 250$; $n > 20$, $np = (250)(0.50) = 125$, $nq = (250)(0.50) = 125$, both np and $nq > 5$

c. $\alpha = 0.05$

Step 4: a. $n = 250$, $x = 113$, $p' = x/n = 113/250 = 0.452$

b. $z = (p' - p)/\sqrt{pq/n}$
$z* = (0.452 - 0.50)/\sqrt{(0.5)(0.5)/250} = -1.52$

c. $P = P(z < -1.52) = P(z > 1.52)$;
Using Table 3, Appendix B, JES2-p521:
$P = 0.5000 - 0.4357 = 0.0643$
Using Table 5, Appendix B, JES2-p523:
$0.0606 < P < 0.0668$
Step 5: a. $P > \alpha$ b. Fail to reject H_O
c. The sample does not provide sufficient evidence to justify the contention that the proportion is less than 0.50, at the 0.05 level of significance.

9.48   Step 1: The proportion, p
Step 2: H_O: p = 0.70 ($\leq$)
H_a: p > 0.70 (higher)
Step 3: a. independence assumed
b. z; n = 300; n > 20, np = (300)(0.70) = 210,
nq = (300)(0.30) = 90, both np and nq > 5
c. $\alpha = 0.05$
d. z(0.05) = 1.65

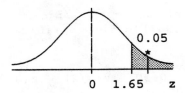

Step 4: a. n = 300, x = 224,   p' = x/n = 224/300 = 0.747
b. $z = (p' - p)/\sqrt{pq/n}$
$z* = (0.747 - 0.70)/\sqrt{(0.7)(0.3)/300} = 1.78$
Step 5: a. z* falls in the critical region, see Step 3d
b. Reject H_O
c. The sample does provide sufficient evidence to justify the contention that the proportion is greater than 0.70, at the 0.05 level of significance.

9.49   The 90% confidence interval for Illustration 9-10 for p was $0.052 < p < 0.118$. 15%, the claimed p, is not contained within this 90% confidence interval.

9.50 a. $\sqrt{p*q*/n} = \sqrt{(0.5)(0.5)/751} = 0.018245 = \underline{0.018}$

b. $E = z(\alpha/2) \cdot \sqrt{p*q*/n}$

   $0.036 = z(\alpha/2)\sqrt{(0.5)(0.5)/721}$

   $0.036 = z(\alpha/2)(0.01862)$

   $1.93 = z(\alpha/2); \quad 0.5000 - 0.4732 = 0.0268$

   $1.93 = z(0.0268)$

   $\alpha/2 = 0.0268$

   $\alpha = 0.0536$

   $1 - \alpha = 1 - 0.0536 = \underline{0.9464}$

SECTION 9.2 EXERCISES

p' = sample proportion	p' = x/n
x = number of successes	
n = sample size (number of independent trials)	

9.51 a. x = number of successes = 45 (with only two outcomes, "success' and "failure")

   n = sample size = number of independent trials = 150

b. p' = 45/150 = # of successes/# of trials = 0.30

   p' = sample proportion of success

9.53 a. Only two possible outcomes, therefore the "opposite" or "not success" is equal to failure. p + q cover the entire outcome list of possibilities for the experiment. p + q = 1

b. p + q = 1 is equivalent to q = 1 - p

c. q = 1 - p = 1 - 0.6 = $\underline{0.4}$

d. p' = 1 - q' = 1 - 0.273 = $\underline{0.727}$

Estimating p - the population proportion

1. point estimate: $p' = x/n$

2. confidence interval: $p' \pm \underbrace{z(\alpha/2) \cdot \sqrt{p'q'/n}}$

                  ↑   maximum error

           point   of estimate
          estimate

9.55 a. $p' = x/n = 75/350 = 0.21$

     b. $E = z(\alpha/2) \cdot \sqrt{p'q'/n} = 1.96\sqrt{(0.21)(0.79)/350}$
          $= (1.96)(0.022) = \underline{0.043}$

Review: JES2-p315, ST-p175, "The Confidence Interval: A Four-Step
Procedure" if necessary.

9.57 Step 1: The proportion of students that support the proposed
               budget amount
       Step 2: a. The sample was randomly selected and each
                 subject's response was independent of those of
                 the others surveyed.
              b. $n = 60$; $n > 20$, $np = (60)(22/60) = 22$,
                 $nq = (60)(38/60) = 38$, np and nq both > 5
              c. $1 - \alpha = 0.99$
       Step 3: a. $n = 60$, $x = 22$
              b. $p' = x/n = 22/60 = 0.367$
       Step 4: a. $z(\alpha/2) = z(0.005) = 2.58$
              b. $E = z(\alpha/2) \cdot \sqrt{p'q'/n} = 2.58\sqrt{(0.367)(0.633)/60}$
                 $= (2.58)(0.0622) = 0.161$
              c. $p' \pm E = 0.367 \pm 0.161$
                   $\underline{0.206 \text{ to } 0.528}$, the 0.99 interval for
                       $p = P(\text{favor budget})$

9.59 $E = z(\alpha/2) \cdot \sqrt{p'q'/n} = z(0.005)\sqrt{(0.51)(0.49)/7225}$
               $= 2.58\sqrt{(0.51)(0.49)/7225} = \underline{0.015}$

9.61 Formula: $p' \pm z(\alpha/2) \cdot \sqrt{p'q'/n}$

    a. $0.3 \pm 1.65\sqrt{(0.3)(0.7)/30} = 0.300 \pm 0.138$

         lower limit = <u>0.162</u>, upper limit = <u>0.438</u>

    b. $0.7 \pm 1.65\sqrt{(0.7)(0.3)/30} = 0.700 \pm 0.138$

         lower limit = <u>0.562</u>, upper limit = <u>0.838</u>

    c. $0.5 \pm 1.65\sqrt{(0.5)(0.5)/10} = 0.500 \pm 0.261$

         lower limit = <u>0.239</u>, upper limit = <u>0.761</u>

    d. $0.5 \pm 1.65\sqrt{(0.5)(0.5)/100} = 0.500 \pm 0.0825$

         lower limit = <u>0.418</u>, upper limit = <u>0.582</u>

    e. $0.5 \pm 1.65\sqrt{(0.5)(0.5)/1000} = 0.500 \pm 0.026$

         lower limit = <u>0.474</u>, upper limit = <u>0.526</u>

    f.   The two answers are symmetric about 0.5.

    g.   As sample size increased, interval became narrower.

9.63   a. $p = P(\text{head}) = 12,012/24,000 = \underline{0.5005}$

    b. $\sqrt{p'q'/n} = \sqrt{(0.5005)(0.4995)/24000} = \underline{0.003227}$

    c. $E = z(\alpha/2) \cdot \sqrt{p'q'/n} = (1.96)(0.003227) = 0.006325 = 0.0063$

       $p \pm E = 0.5005 \pm 0.0063$

         <u>0.4942 to 0.5068</u>, the 0.95 interval for $p = P(\text{head})$

    d. - f. Each student will have different results. Each set of results will yield an empirical probability whose value is very close to 0.50; in fact, you should expect 95% of such results to be within 0.0063 of 0.50.

$$
\boxed{
\begin{array}{c}
\text{Sample Size Determination Formula for a Population Proportion} \\[4pt]
n = \dfrac{[z(\alpha / 2)]^2 \cdot p^* \cdot q^*}{E^2} \\[6pt]
\textbf{Sample Size - A Four-Step Procedure} \\[4pt]
\end{array}
}
$$

Step 1: Use the level of confidence, $1-\alpha$, to find $z(\alpha/2)$
Step 2: Find the maximum error of estimate
Step 3: Determine p^* and $q^* = 1 - p^*$ (if not given, use $p^* = 0.5$)
Step 4: Use formula to find n

9.65 a. Step 1: $1 - \alpha = 0.90$; $z(\alpha/2) = z(0.05) = 1.65$
   Step 2: $E = 0.02$
   Step 3: $p^* = 0.16$ and $q^* = 0.84$
   Step 4: $n = \{[z(\alpha/2)]^2 \cdot p^* \cdot q^*\}/E^2$
        $= (1.65^2)(0.16)(0.84)/(0.02^2) = 914.76 = \underline{915}$

   b. Step 1: $1 - \alpha = 0.90$; $z(\alpha/2) = z(0.05) = 1.65$
   Step 2: $E = 0.04$
   Step 3: $p^* = 0.16$ and $q^* = 0.84$
   Step 4: $n = \{[z(\alpha/2)]^2 \cdot p^* \cdot q^*\}/E^2$
        $= (1.65^2)(0.16)(0.84)/(0.04^2) = 228.69 = \underline{229}$

   c. Step 1: $1 - \alpha = 0.98$; $z(\alpha/2) = z(0.01) = 2.33$
   Step 2: $E = 0.02$
   Step 3: $p^* = 0.16$ and $q^* = 0.84$
   Step 4: $n = \{[z(\alpha/2)]^2 \cdot p^* \cdot q^*\}/E^2$
        $= (2.33^2)(0.16)(0.84)/(0.02^2) = 1824.11 = \underline{1825}$

   d. Increasing the maximum error decreases the sample size.

   e. Increasing the level of confidence increases the sample size.

9.67 Step 1: $1 - \alpha = 0.95$; $z(\alpha/2) = z(0.025) = 1.96$
   Step 2: $E = 0.03$
   Step 3: $p^* = 0.5$ and $q^* = 0.5$
   Step 4: $n = \{[z(\alpha/2)]^2 \cdot p^* \cdot q^*\}/E^2$
        $= (1.96^2)(0.5)(0.5)/(0.03^2) = \underline{1068}$

Hypotheses are written with the same rules as before. Now replace μ, the population mean, with p, the population proportion.

(ex.: H_O: p = P(driving a convertible) = 0.45 vs.
    H_a: p = P(driving a convertible) $\neq$ 0.45,
if driving a convertible is considered the success)

Review: JES2-pp335&337, ST-p184, if necessary.

9.69 a. H_O: p = P(work) = 0.60 ($\leq$)   vs. H_a: p > 0.60

b. H_O: p = P(win tonight) = 0.50 ($\geq$)   vs. H_a: p < 0.50

c. H_O: p = P(interested in quitting) = 1/3 ($\leq$)
       vs. H_a: p > 1/3

d. H_O: p = P(believe in spanking) = 0.50 ($\geq$)
       vs. H_a: p < 0.50

e. H_O: p = P(vote for) = 0.50 ($\leq$)   vs. H_a: p > 0.50

f. H_O: p = P(seriously damaged) = 3/4 ($\geq$)
       vs. H_a: p < 3/4

g. H_O: p = P(H|tossed fairly) = 0.50   vs. H_a: p $\neq$ 0.50

h. H_O: p = P(odd|random) = 0.50   vs. H_a: p $\neq$ 0.50

Determining the p-value will follow the same procedures as before. We are again working with the normal distribution. Review: JES2-p341, Table 8-6; ST-p187, if necessary.

9.71 a. $\mathbf{P}$ = 2P(z > 1.48) = 2(0.5000 - 0.4306) = 2(0.0694) = <u>0.1388</u>

b. $\mathbf{P}$ = 2P(z < -2.26) = 2P(z > 2.26) = 2(0.5000 - 0.4881)
       = 2(0.0119) = <u>0.0238</u>

c. $\mathbf{P}$ = P(z > 0.98) = (0.5000 - 0.3365) = <u>0.1635</u>

d. $\mathbf{P}$ = P(z < -1.59) = P(z > 1.59) = (0.5000 - 0.4441)
       = <u>0.0559</u>

Since n ≤ 15 and **x** is discrete, Table 2 (Appendix B, JES2-p518) will be used to determine the level of significance, α.
x can be any value, 0 through n, for each experiment.
List all values in numerical sequence.
Draw a vertical line separating the set of values that belong in the critical region and the set of values that belong in the noncritical region.
Add all of the probabilities associated with those numbers in the critical region to find α.

9.73 a. $\alpha = P[x = 12, 13, 14, 15 | B(n=15, p=0.5)]$
   $= 0.014 + 0.003 + 2(0+) = \underline{0.017}$

b. $\alpha = P[x = 0, 1 | B(n=12, p=0.3)]$
   $= 0.014 + 0.071 = \underline{0.085}$

c. $\alpha = P[x = 0, 1, 2, 3, 9, 10 | B(n=10, p=0.6)]$
   $= (0+) + 0.002 + 0.011 + 0.042 + 0.040 + 0.006 = \underline{0.101}$

d. $\alpha = P[x = 4, 5, 6, \ldots, 14 | B(n=14, p=0.05)]$
   $= 0.004 + 10(0+) = \underline{0.004}$

9.75 a. Correctly fail to reject H_O

b. $\alpha = P[x = 14, 15 | B(n = 15, p = 0.7)]$
   $= 0.031 + 0.005 = \underline{0.036}$

c. Commit a type II error

d. **P** $= P[x = 13, 14, 15 | B(n = 15, p = 0.7)]$
   $= 0.092 + 0.031 + 0.005 = \underline{0.128}$

Review: "The Probability-Value Hypothesis Test: A Five-Step Procedure"; JES2-p334, ST-p183 and/or "The Classical Hypothesis Test: A Five-Step Procedure"; JES2-p350, ST-p192, if necessary.

Use $z = \dfrac{p' - p}{\sqrt{\dfrac{pq}{n}}}$, for calculating the test statistic.

p' = x/n, if not given directly.

<u>Hint for writing the hypotheses for exercise 9.77</u>
Look at the last sentence in the exercise, "If the consumer group ... that <u>fewer than</u> 90% ...?" The words "fewer than" indicate <u>less than</u> (<), therefore the alternative is less than (<). The negation is "NOT less than", which is > or =. Equality (=) is used in the null hypothesis, but stands for greater than or equal to (≥). Remember to use *p* as the population parameter in the hypotheses.

9.77 a. P-value approach:

Step 1: The proportion of claims settled within 30 days
Step 2: H_O: p = P(claim is settled within 30 days) = 0.90 (≥)
$\qquad$ H_a: p < 0.90
Step 3: a. independence assumed
$\qquad$ b. z; n = 75; n > 20, np = (75)(0.90) = 67.5,
$\qquad\quad$ nq = (75)(0.10) = 7.5, both np and nq > 5
$\qquad$ c. α = 0.05
Step 4: a. n = 75, x = 55, p' = x/n = 55/75 = 0.733
$\qquad$ b. z = (p' - p)/$\sqrt{pq/n}$
$\qquad\quad$ z* = (0.733 - 0.900)/$\sqrt{(0.9)(0.1)/75}$ = -4.82
$\qquad$ c. **P** = P(z < -4.82) = P(z > 4.82);
$\qquad\qquad$ Using Table 3, Appendix B, JES2-p521:
$\qquad\qquad$ **P** = 0.5000 - 0.499997 = 0.000003
$\qquad\qquad$ Using Table 5, Appendix B, JES2-p523:
$\qquad\qquad$ **P** > 0+
Step 5: a. **P** < α $\qquad$ b. Reject H_O
$\qquad$ c. The sample provides sufficient evidence that p is
$\qquad\quad$ significantly less than 0.90; it appears that
$\qquad\quad$ less than 90% are settled within 30 days as
$\qquad\quad$ claimed, at the 0.05 level of significance.

b. Classical approach:

Step 1: The proportion of claims settled within 30 days

Step 2: H_O: p =P(claim is settled within 30 days) = 0.90 ($\geq$)
H_a: p < 0.90

Step 3: a. independence assumed
b. z; n = 75; n > 20, np = (75)(0.90) = 67.5,
nq = (75)(0.10) = 7.5, both np and nq > 5
c. α = 0.05
d. -z(0.05) = -1.65

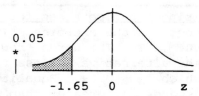

0.05

-1.65   0   z

Step 4: a. n = 75, x = 55, p' = x/n = 55/75 = 0.733
b. z = (p' - p)/$\sqrt{pq/n}$
z* = (0.733 - 0.900)/$\sqrt{(0.9)(0.1)/75}$ = -4.82

Step 5: a. z* falls in the critical region, see Step 3d
b. Reject H_O
c. The sample provides sufficient evidence that p is
significantly less than 0.90; it appears that
less than 90% are settled within 30 days as
claimed.

Hint for writing the hypothesis for exercise 9.79

Look at the first sentence of the exercise, "A politician claims she
will receive 60% of the vote ..." The words "will receive" indicate
equality (=). Since a politician would be interested in a majority,
anything equal to or greater than the stated percentage would be
acceptable. Greater than or equal to ($\geq$) includes the equality, therefore it belongs in the null hypothesis. Continue to write the null
hypothesis with the equal sign (=), but include the greater-than or
equal-to sign($\geq$) in parentheses after it. The negation is "...less
than," hence the alternative hypothesis must have a less-than sign(<).

9.79 a. P-value approach:

Step 1: The proportion of vote for a politician in an
upcoming election

Step 2: H_O: p = P(vote for) = 0.60
[will receive 60% of vote] ($\geq$)
H_a: p < 0.60 [will receive less than 60%]

Step 3: a. independence assumed
b. z; n = 100; n > 20, np = (100)(0.60) = 60,
nq = (100)(0.40) = 40, both np and nq > 5
c. α = 0.05
Step 4: a. n = 100, x = 50, p' = x/n = 50/100 = 0.500
b. z = (p' - p)/$\sqrt{pq/n}$
z* = (0.500 - 0.600)/$\sqrt{(0.6)(0.4)/100}$ = -2.04
c. P = P(z < -2.04) = P(z > 2.04);
Using Table 3, Appendix B, JES2-p521:
P = 0.5000 - 0.4793 = 0.0207
Using Table 5, Appendix B, JES2-p523:
0.0202 < P < 0.0228
Step 5: a. P < α b. Reject H_O
c. The sample provides sufficient evidence that the
proportion is significantly less than 0.60, at
the 0.05 level; it appears that less than 60%
support her.

b. Classical approach:

Step 1: The proportion of vote for a politician in an
upcoming election
Step 2: H_O: p = P(vote for) = 0.60
[will receive 60% of vote] ($\geq$)
H_a: p < 0.60 [will receive less than 60%]
Step 3: a. independence assumed
b. z; n = 100;   n > 20, np = (100)(0.60) = 60,
nq = (100)(0.40) = 40, both np and nq > 5
c. α = 0.05
d. -z(0.05) = -1.65

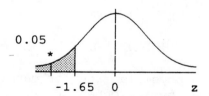

Step 4: a. n = 100, x = 50, p' = x/n = 50/100 = 0.500
b. z = (p' - p)/$\sqrt{pq/n}$
z* = (0.500 - 0.600)/$\sqrt{(0.6)(0.4)/100}$ = -2.04
Step 5: a. z* falls in the critical region, see Step 3d
b. Reject H_O

c. The sample provides sufficient evidence that the proportion is significantly less than 0.60, at the 0.05 level; it appears that less than 60% support her.

9.81 P-value approach:

Step 1: The proportion of adults worried about maintaining mortgage payments

Step 2: H_O: p = P(worried about mortgage payments) = 0.27 ($\geq$)
H_a: p < 0.27

Step 3: a. independence assumed
b. z; n = 759; n > 20, np = (759)(0.27) = 204.93, nq = (759)(0.73) = 554.07, both np and nq > 5
c. α = 0.05

Step 4: a. n = 759, x = 150, p' = x/n = 150/759 = 0.198
b. z = (p' - p)/$\sqrt{pq/n}$
$z*$ = (0.198 - 0.270)/$\sqrt{(0.27)(0.73)/759}$ = -4.47
c. **P** = P(z < -4.47) = P(z > 4.47);
Using Table 3, Appendix B, JES2-p521:
P = 0.5000 - 0.499997 = 0.000003
Using Table 5, Appendix B, JES2-p523:
P = 0+

Step 5: a. **P** < α b. Reject H_O
c. The sample provides sufficient evidence that the proportion is significantly less than 0.27, at the 0.05 level; it appears that less than 27% are worried about maintaining their mortgage payments.

CHAPTER EXERCISES

9.83 Step 1: The mean physician fee for cataract removal
Step 2: a. normality indicated
b. t   c. 1-α = 0.99
Step 3: a. n = 25, $\overline{x}$ = 1550,   s = 125
b. $\overline{x}$ = 1550
Step 4: a. α/2 = 0.01/2 = 0.005; df = 24; t(24, 0.005) = 2.80
b. E = t(df,α/2)·(s/$\sqrt{n}$) = (2.80)(125/$\sqrt{25}$) = 70
c. $\overline{x}$ ± E = 1550 ± 70
$\underline{\$1,480 \text{ to } \$1,620}$, the 0.99 estimate for μ

9.85 $t = (\overline{x} - \mu)/(s/\sqrt{n})$

$t* = (6300 - 6700)/(3250/\sqrt{500}) = -2.75$

$P = P(t < -2.75|df = 499) = P(t > 2.75)|df = 499);$
   Using Table 6, Appendix B, JES2-p524:
      $P < 0.005$
   Using Table 7, Appendix B, JES2-p525:
      $0.004 < P < 0.005$

9.87 Summary of data: n = 12, $\Sigma x = 41.3$, $\Sigma x^2 = 146.83$

$\overline{x} = \Sigma x/n = 41.3/12 = \underline{3.44}$

$s^2 = [\Sigma x^2 - (\Sigma x)^2/n]/(n - 1)$

$\quad = [146.83 - (41.3^2/12)]/11 = 0.4263$

$s = \sqrt{s^2} = \sqrt{0.4263} = \underline{0.653}$

Step 1: The mean of this year's pollution readings
Step 2: H_o: $\mu = 3.8$ ($\geq$)
        H_a: $\mu < 3.8$ [lower]
Step 3: a. normality indicated
        b. t   c. $\alpha = 0.05$
Step 4: a. n = 12, $\overline{x} = 3.44$,   s = 0.653
        b. $t = (\overline{x} - \mu)/(s/\sqrt{n})$
           $t* = (3.44 - 3.8)/(0.653/\sqrt{12}) = -1.91$
        c. $P = P(t < -1.91|df = 11) = P(t > 1.91|df = 11);$
           Using Table 6, Appendix B, JES2-p524:
           $0.025 < P < 0.05$
           Using Table 7, Appendix B, JES2-p525:
           $0.034 < P < 0.043$
Step 5: a. $P < \alpha$ b. Reject H_o
        c. The sample does provide sufficient evidence to
           justify the contention that the mean of this
           year's pollution readings is significantly lower
           than last year's mean, at the 0.05 level.

9.89 Step 1: The mean for a standardized reading test used in
                Nebraska
        Step 2: H_o: $\mu = 80$ [same as state]
                H_a: $\mu \neq 80$ [different]

```
Step 3: a. normality indicated
        b. t            c. α = 0.05
Step 4: a. n = 20, x̄ = 77.5,  s = 2.5
        b. t = (x̄ - μ)/(s/√n)
           t* = (77.5 - 80.0)/(2.5/√20) = -4.47
```

```
     a.    Step 4c. P = 2P(t < -4.47|df = 19)
                      = 2P(t > 4.47|df = 19);
                      Using Table 6, Appendix B, JES2-p524:
                      P < 2(0.005); P < 0.01
                      Using Table 7, Appendix B, JES2-p525:
                      P = 0+
           Step 5a. P < α
     OR
     b.    Step 3d. ±t(19, 0.025) = ±2.09
```

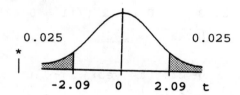

```
           Step 5a. t* falls in the critical region,
                    see * Step 3d
```

```
Step 5: b. Reject H_O
        c. The sample does provide sufficient evidence to
           show the mean is significantly different than 80,
           at the 0.05 level.
```

9.91 a. n = 800;   trial = one person surveyed;
         success = doctor should not be prosecuted;
         p = P(success);
         x = n(success) = 0, 1, 2, ... , 800

      b. It is a statistic, p' = 2/3, since it describes the
         results of the sample. The value was obtained by
         dividing the number who said "doctor should not be
         prosecuted" by the number of people polled (800).

c. $E = z(\alpha/2) \cdot \sqrt{p'q'/n}$

$= 1.96\sqrt{(2/3)(1/3)/800} = \underline{0.033}$

d. Rounded to the nearest half of 1%, it is the same.

9.93 Step 1: The proportion who recognize and respect a particular woman golfer

Step 2: a. The sample was randomly selected and each subject's response was independent of those of the others surveyed.

b. $n = 100$; $n > 20$, $np = (100)(16/100) = 16$, $nq = (100)(84/100) = 84$, np and nq both > 5

c. $1 - \alpha = 0.95$

Step 3: a. $n = 100$, $x = 16$

b. $p' = x/n = 16/100 = 0.16$

Step 4: a. $z(\alpha/2) = z(0.025) = 1.96$

b. $E = z(\alpha/2) \cdot \sqrt{p'q'/n} = 1.96\sqrt{(0.16)(0.84)/100}$
$= (1.96)(0.03666) = 0.072$

c. $p' \pm E = 0.160 \pm 0.072$

$\underline{0.088 \text{ to } 0.232}$, the 0.95 interval for $p =$ P(recognized)

9.95

p	= 0.1	0.2	0.3	0.4	0.5	0.6	0.7	0.8	0.9
q	= 0.9	0.8	0.7	0.6	0.5	0.4	0.3	0.2	0.1
pq	= 0.09	0.16	0.21	0.24	0.25	0.24	0.21	0.16	0.09

9.97 a. $E = z(\alpha/2) \cdot \sqrt{p \cdot q/n} = 1.96\sqrt{(0.64)(0.36)/1254} = \underline{0.027}$

b. To the nearest percent, they are equal.

c. $n = [(z(\alpha/2))^2 \cdot p \cdot q]/E^2$

$= [(1.96^2)(0.64)(0.36)]/(0.02^2) = \underline{2213}$

9.99 Let x = number of defective in the sample of 50. When H_O is true, we may treat x as a binomial variable with n = 50 and p = 0.005.

α is **P**(of rejecting H_O when it is true):

$\alpha = P[x \geq 2 | B(n = 50, p = 0.005)]$

$= 1.0 - P[x = 0 \text{ or } 1 | B(n = 50, p = 0.005)]$

$= 1.0 - [\binom{50}{0} \cdot (0.005)^0 \cdot (0.995)^{50} + \binom{50}{1} \cdot (0.005)^1 \cdot (0.995)^{49}]$

$= 1.0000 - [0.7783 + 0.1956] = \underline{0.0261}$

9.101 H_O: p = P(7 or 11 picked randomly) = 2/15

H_a: p ≠ 2/15 [not picked randomly]

$\alpha = 0.05$

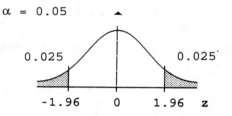

0.025   0.025

-1.96   0   1.96   z

$z = (p' - p) / \sqrt{pq/n}$

$1.96 = [p' - (2/15)] / \sqrt{(2/15)(13/15)/54}$

$p' = (2/15) + (1.96)\sqrt{(2/15)(13/15)/54} = 0.224$

Since p' = x/n; then x = n·p' = 54(0.224) = 12.09 = $\underline{13}$

CHAPTER 10 ∇ INFERENCES INVOLVING TWO POPULATIONS

Chapter Preview

In Chapters 8 and 9, the concepts of confidence intervals and hypothesis tests were introduced. Each of these was demonstrated with respect to a single mean or proportion. In Chapter 10, these concepts will be extended to include two means or two proportions, thereby enabling us to compare two populations. Distinctions will have to be made with respect to dependent and independent samples, in order to select the appropriate testing procedure and test statistic.

SECTION 10.1 MARGIN EXERCISES

10.1 Divide the class into two groups, males and females. Randomly select a sample from each of the two groups.

10.2 Randomly select a set of students, obtaining the two heights from each of the selected students.

10.3 Identical twins are so much alike that the information obtained from one would not be independent from the information obtained from the other twin.

SECTION 10.1 EXERCISES

INDEPENDENT SAMPLES - Two samples are independent if the selection of one sample from a population has no effect on the selection of the other sample from another population. (They do not have to be different populations.)
ex.: the repair costs for two different brands of VCRs

DEPENDENT SAMPLES (paired samples) - two samples are dependent if the objects or individuals selected for one sample from a population are paired in some meaningful way with the objects or individuals selected for the second sample from the same or another population.
ex.: "before and after" experiments - change in weight for smokers who became nonsmokers

10.5   Dependent samples. The two sets of data were obtained from
the same set of 20 people, each person providing one piece of
data for each sample.

10.7   Independent samples. The gallon of paint serves as the
population of many (probably millions) particles. Each set
of 10 specimens forms a separate and independent sample.

10.9   a. Independent samples will result if the two sets are
selected in such a way that there is no relationship
between the two resulting sets.

   b. Dependent samples will result if the 1,000 men and women
were husband and wife or if they were brother and sister,
or related in some way.

SECTION 10.2 MARGIN EXERCISES

10.11

Pairs	1	2	3	4	5
d = A - B	1	1	0	2	-1

10.12 $n = 5$, $\Sigma d = 3$, $\Sigma d^2 = 7$

$\overline{d} = \Sigma d/n = 3/5 = \underline{0.6}$

$s_d = \sqrt{(\Sigma d^2 - (\Sigma d)^2/n)/(n-1)} = \sqrt{(7 - (3)^2/5)/(4)}$
$= \sqrt{1.3} = \underline{1.14}$

10.13 $t(15, 0.025) = \underline{2.13}$; it determines the number of standard
errors the confidence interval should extend in either
direction from the mean difference. Its value depends on
the sample size and level of confidence.

10.14 a. Step 1: The mean difference, μ_d
   Step 2: a. normality assumed
          b. t   c. $1-\alpha = 0.95$
   Step 3: a. $n = 26$, $\bar{d} = 6.3$, $s_d = 5.1$
          b. $\bar{d} = 6.3$
   Step 4: a. $\alpha/2 = 0.05/2 = 0.025$; $df = 25$;
             $t(25, 0.025) = 2.06$
          b. $E = t(df, \alpha/2) \cdot (s_d/\sqrt{n}) = (2.06)(5.1/\sqrt{26})$
             $= (2.06)(1.0002) = 2.06$
          c. $\bar{d} \pm E = 6.3 \pm 2.06$
          <u>4.24 to 8.36</u>, the 0.95 confidence interval for μ_d

b. The same $\bar{d}$ and s_d values were used with a much larger n, resulting in a narrower confidence interval.

10.15 With the before data in C1 and the after data in C2, MINITAB commands, LET C3 = C1 - C2 and TINT 95 C3 result in:

Confidence Intervals

Variable	N	Mean	StDev	SEMean	95% C.I.
C3	6	3.0	3.35	1.37	(-0.51,6.51)

10.16 Data Summary: $n = 5$, $\Sigma d = 90$, $\Sigma d^2 = 2700$

   Step 1: The mean difference, μ_d
   Step 2: $H_O: \mu_d = 0$ ($\leq$)
          $H_a: \mu_d > 0$
   Step 3: a. normality indicated
          b. t   c. $\alpha = 0.05$
   Step 4: a. $n = 5$, $\bar{d} = 18$, $s_d = 16.43$
          b. $t^* = (\bar{d} - \mu_d)/(s_d/\sqrt{n})$
             $= (18 - 0)/(16.43/\sqrt{5}) = 2.45$
          c. $P = P(t^* > 2.45 | df = 4)$;
             Using Table 6, Appendix B, JES2-p524:
             $0.025 < P < 0.05$
             Using Table 7, Appendix B, JES-p525:
             $0.033 < P < 0.037$
   Step 5: a. $P < \alpha$ b. Reject H_O
          c. At the 0.05 level of significance, there is sufficient evidence that the mean difference is greater than zero.

10.17 With M data in C1 and N data in C2, MINITAB commands,
LET C3 = C1-C2   and   TTESt 0 C3; ALTErnative -1.
results in:

T-Test of the Mean
Test of mu = 0.00 vs   mu < 0.00

Variable	N	Mean	StDev	SEMean	T	P-value
C3	6	-2.33	4.41	1.80	-1.30	0.13

With α = 0.02, p > α, fail to reject.

10.18 Data Summary: n = 5, Σd = 12, Σd^2 = 92

Step 1: The mean difference, μ_d
Step 2: H_o: μ_d = 0 ($\leq$)
          H_a: μ_d > 0
Step 3: a. normality indicated
        b. t   c. α = 0.01
        d. t(4, 0.01) = 3.75

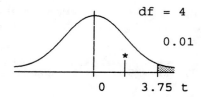

Step 4: a. n = 5, $\bar{d}$ = 2.4, s_d = 3.97
        b. t* = $(\bar{d} - \mu_d)/(s_d/\sqrt{n})$
             = $(2.4 - 0)/(3.97/\sqrt{5})$ = 1.35
Step 5: a. t* falls in the noncritical region, see Step 3d
        b. Fail to reject H_o
        c. At the 0.01 level of significance, there is not
           sufficient evidence that the mean difference is
           greater than zero.

SECTION 10.2 EXERCISES

Estimating μ_d - the population mean difference

1. point estimate: $\bar{d} = \dfrac{\Sigma d}{n}$

2. confidence interval: $\bar{d} \pm t(df,\ \alpha/2) \cdot (s_d/\sqrt{n})$

    ↑   ↑   ↑

   point   confidence   estimated

   estimate   coefficient   standard error

maximum error of estimate

Follow the steps outlined in "The Confidence Interval: A Four-Step Procedure" in: JES2-p315, ST-p175.

10.19 Data Summary: n = 8, Σd = 8, Σd^2 = 48

a. Point estimate = 1.0

b. Step 1: The mean reduction in diastolic blood pressure following a two week salt-free diet (d = B - A)
Step 2: a. normality indicated
b. t   c. 1-α = 0.98
Step 3: a. n = 8, $\bar{d}$ = 1, s_d = 2.39
b. $\bar{d}$ = 1
Step 4: a. $\alpha/2$ = 0.02/2 = 0.01; df = 7;
t(7, 0.01) = 3.00
b. E = $t(df,\alpha/2)\cdot(s_d/\sqrt{n})$ = (3.00)(2.39/$\sqrt{8}$)
= (3.00)(0.845) = 2.53
c. $\bar{d} \pm E$ = 1 ± 2.53
-1.53 to 3.53, the 0.98 confidence interval for μ_d

10.21 Sample statistics: d = A - B

n = 8, $\bar{d}$ = 3.75, s_d = 5.726

Step 1: The mean difference in weight gain for pigs fed ration A as compared to those fed ration B
Step 2: a. normality indicated
b. t   c. 1-α = 0.95

Step 3: a. $n = 8$, $\bar{d} = 3.75$, $s_d = 5.726$
   b. $\bar{d} = 3.75$
Step 4: a. $\alpha/2 = 0.05/2 = 0.025$; $df = 7$; $t(7, 0.025) = 2.36$
   b. $E = t(df, \alpha/2) \cdot (s_d/\sqrt{n}) = (2.36)(5.726/\sqrt{8})$
      $= (2.36)(2.0244) = 4.78$
   c. $\bar{d} \pm E = 3.75 \pm 4.78$
      <u>-1.03 to 8.53</u>, the 0.95 interval for μ_d

WRITING HYPOTHESES FOR TEST OF TWO DEPENDENT MEANS

μ_d = population mean difference

<u>null hypothesis</u> - H_O: $\mu_d = 0$
   ("the mean difference equals 0, that is, there is
      no difference within the pairs of data")

<u>possible alternative hypotheses</u> -
   H_a: $\mu_d > 0$
   H_a: $\mu_d < 0$
   H_a: $\mu_d \neq 0$,
   ("the mean difference is significant, that is,
      there is a difference within the pairs of data")

10.23 a. H_O: $\mu_d = 10$ ($\leq$); H_a: $\mu_d > 10$; d = posttest - pretest

   b. H_O: $\mu_d = 10$ ($\geq$); H_a: $\mu_d < 10$; d = after - before

   c. H_O: $\mu_d = 12$ ($\geq$); H_a: $\mu_d < 12$; d = before - after

   d. H_O: $\mu_d = 0$; H_a: $\mu_d \neq 0$; d = after - before

---

Hypothesis Tests for Two Dependent Means

In this form of hypothesis test, each data value of the first
sample is compared to its corresponding (or paired) data value in
the second sample. The differences between these paired data
values are calculated, thereby forming a sample of differences or d
values. It is these differences or d values that we wish to use to
test the difference between two dependent means.

...

Review the parts to a hypothesis test (p-value & classical) as outlined in: JES2-pp334&350, ST-pp183&192, if needed. Changes will occur in:

1) the calculated value of the test statistic, t;

$$t = \frac{\overline{d} - \mu_d}{s_d / \sqrt{n}} \quad , \quad \text{where } \overline{d} = \frac{\Sigma\, d}{n} \quad , s_d = \sqrt{\frac{\Sigma\, d^2 - (\Sigma\, d)^2 / n}{n - 1}}$$

and n = # of paired differences

2) a. p-value approach
   Use Table 6 (Appendix B, JES2-p524) to <u>estimate</u> the p-value
   1) Locate df row.
   2) Locate the absolute value of the calculated t-value between two critical values in the df row.
   3) The p-value is in the interval between the two corresponding probabilities at the top of the columns; read the bounds from the *one-tailed* heading if H_a is one-tailed, or from the *two-tailed* headings if H_a is two-tailed.

   OR
   Use Table 7 (Appendix B, JES2-p525) to estimate or place bounds on the p-value
   1) locate the absolute value of the calculated t-value along the left margin and the df along the top, then read the p-value directly from the table where the row and column intersect

   OR
   2) locate the absolute value of the calculated t-value and its df between appropriate bounds. From the box formed at the intersection of these row(s) and column(s), use the upper left and lower right values for the bounds on **P**.

   b. classical approach
   Use Table 6 (Appendix B, JES2-p524) with df = n - 1
   and α to find the critical value

   . . .

-- 243 --

3) if H_O is rejected, a significant difference as stated in H_a is indicated
   if H_O is not rejected, no significant difference is indicated

NOTE: To find $\bar{d}$ and s_d, set up a table of corresponding pairs of data. Calculate d, the difference (be careful to subtract in the same direction each time). Calculate a d^2 for each pair and find summations, Σd and $\Sigma(d^2)$.

The sample of paired differences are assumed to be selected from an approximately normally distributed population with a mean μ_d and a standard deviation σ_d. Since σ_d is unknown, the calculated t-statistic is found using an estimated standard error of $s_d/\sqrt{n}$.

<u>Hint for writing the hypotheses for exercise 10.25</u>
Look at the second and third sentences of the exercise; "The data ... where d is the amount of corrosion on the coated portion subtracted from the amount of corrosion on the uncoated portion. Does this sample provide sufficient reason to conclude that the coating is <u>beneficial</u>?" For the coating to be beneficial, there would have to be less corrosion on the coated portion. This implies that the differences, "uncoated - coated," if beneficial, would be <u>positive</u>, which in turn implies a <u>greater than zero</u> (> 0). Therefore the alternative hypothesis is a greater than (>). The negation is "NOT beneficial", which indicates < or =. Therefore the null hypothesis should be written with the equal sign (=), but include the less-than or equal-to sign ($\leq$) in parentheses after it.

10.25 Sample statistics: $n = 40$, $\bar{d} = 5.5$, $s_d = 11.34$

a. Step 1: The mean difference between coated and uncoated sections of steel pipe

Step 2: H_O: $\mu_d = 0$ ($\leq$)
   H_a: $\mu_d > 0$ (beneficial)

Step 3: a. normality assumed, CLT with $n = 40$.
   b. t   c. $\alpha = 0.01$

Step 4: a. $n = 40$, $\bar{d} = 5.5$, $s_d = 11.34$
   b. $t* = (\bar{d} - \mu_d)/(s_d/\sqrt{n})$
     $= (5.5 - 0.0)/(11.34/\sqrt{40}) = 3.067$
   c. **P** $= P(t > 3.067 | df = 39)$;
     Using Table 6, Appendix B, JES2-p524:
     P < 0.005
     Using Table 7, Appendix B, JES2-p525:
     P ≈ 0.002

Step 5: a. **P** < α   b. Reject H_o
        c. At the 0.01 level of significance, there is a
           significant benefit to coating the pipe.

Reviewing how to determine the test criteria in: JES2-pp354&355,
ST-p193&212, may be helpful.

b. Step 1: The mean difference between coated and uncoated
           sections of steel pipe
   Step 2: H_o: $\mu_d = 0$ ($\le$)
           H_a: $\mu_d > 0$ (beneficial)
   Step 3: a. normality assumed, CLT with n = 40.
           b. t   c. α = 0.01
           d. t(39, 0.01) = 2.44

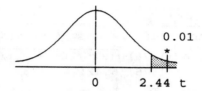

$$0.01$$
$$0 \qquad 2.44 \; t$$

   Step 4: a. n = 40, $\bar{d}$ = 5.5, s_d = 11.34
           b. $t* = (\bar{d} - \mu_d)/(s_d/\sqrt{n})$
              $= (5.5 - 0.0)/(11.34/\sqrt{40}) = 3.067$
   Step 5: a. t* falls in the critical region,
              see * Step 3d
           b. Reject H_o
           c. At the 0.01 level of significance, there is a
              significant benefit to coating the pipe.

MINITAB can perform a hypothesis test on paired data using the **LET**
and **TTEST** commands.
The **LET** command, written in the form of an equation, performs a row
by row subtraction between the data values, thereby forming a
column of sample differences.
If the original two sets of data are in C1 and C2, then

        LET C3 = C2 - C1 <u>or</u> **LET** C3 = C1 - C2

could be used. The choice is dependent on the order of subtraction
needed to match the "planned" approach as determined by H_a.
The hypothesis test will be performed on this column of differences
(C3).   Additional information can be found in JES2-p423.

10.27 MINITAB verify - answers given in exercise.

SECTION 10.3 MARGIN EXERCISES

10.29 $\sqrt{(s_1^2 / n_1) + (s_2^2 / n_2)} = \sqrt{(190 / 12) + (150 / 18)} = \sqrt{24.1667} = \underline{4.92}$

10.30 Case I: df will be between 17 and 40
Case II; df = 17 (smaller df)

10.31 Step 1: The difference between two means, $\mu_1 - \mu_2$
Step 2: a. normality indicated
b. t   c. $1 - \alpha = 0.90$
Step 3: a. sample information given in exercise
b. $\overline{x}_1 - \overline{x}_2 = 35 - 30 = 5$
Step 4: a. $\alpha/2 = 0.10/2 = 0.05$; df = 14;
t(14, 0.05) = 1.76

b. $E = t(df, \alpha/2) \cdot \sqrt{(s_1^2 / n_1) + (s_2^2 / n_2)}$
$= (1.76)\sqrt{(22^2/20) + (16^2/15)}$
$= (1.76)(6.42) = 11.3$

c. $(\overline{x}_1 - \overline{x}_2) \pm E = 5 \pm 11.3$
$\underline{-6.3 \text{ to } 16.3}$, the 0.90 confidence interval for $\mu_1 - \mu_2$

10.32 $t^* = [(\overline{x}_2 - \overline{x}_1) - (\mu_2 - \mu_1)] / \sqrt{(s_2^2 / n_2) + (s_1^2 / n_1)}$
$= [(43.1 - 38.2) - 0]/\sqrt{(10.6^2/25) + (14.2^2/18)} = \underline{1.24}$

10.33 $P = 2P(t^* > 1.3 | df = 18)$;
Using Table 6, Appendix B, JES2-p524:
$\underline{0.20 < P < 0.50}$
Using Table 7, Appendix B, JES2-p525:
$P = 2(0.105) = \underline{0.210}$

10.34 With the smaller degrees of freedom, df = 9, a higher calculated value is needed making it more difficult to reject H_O. This is due to the lack of reliability with a small sample.

10.35 a. The samples are taken from two different sets of schools. A school is either AACSB-accredited or not; it can not be both.

b. The probability that there is no difference between the two types of schools is just about 0.

c. df = 73, 0.5(0.394) = 0.197;
   Using Table 6: 0.68 < t* < 1.29
   Using Table 7: 0.8 < t* < 0.9

d. A five-point scale was used and there is no indication of normality given.

SECTION 10.3 EXERCISES

Estimating $(\mu_1 - \mu_2)$ - the difference between two population means,

1. Point Estimate: $\bar{x}_1 - \bar{x}_2$

2. Confidence Interval

$$(\bar{x}_1 - \bar{x}_2) \pm t(df, \alpha/2) \cdot \sqrt{(s_1^2 / n_1) + (s_2^2 / n_2)}$$

↑	↑	↑
point estimate	confidence coefficient	estimated standard error

maximum error of estimate

estimate df by using the smaller value of df_1 or df_2

Review "The Confidence Interval: A Four-Step Procedure" in:
JES2-p315, ST-p175, if necessary.
Subtract sample means ($\bar{x}_1 - \bar{x}_2$ or $\bar{x}_2 - \bar{x}_1$) in whichever order results in a positive difference.
Also, use appropriate subscripts to designate the source.

10.37   Step 1: The difference between the mean absorption rates
   for two drugs, $\mu_B - \mu_A$
   Step 2: a. normality indicated
        b. t   c. $1-\alpha = 0.98$
   Step 3: a. $n_A = 36$, $\bar{x}_A = 7.9$, $s_A = 0.11$
          $n_B = 36$, $\bar{x}_B = 8.5$, $s_B = 0.10$
        b. $\bar{x}_B - \bar{x}_A = 8.5 - 7.9 = 0.60$

Step 4: a. $\alpha/2 = 0.02/2 = 0.01$; df = 35;
   $t(35, 0.01) = 2.44$

b. $E = t(df, \alpha/2) \cdot \sqrt{(s_B^2 / n_B) + (s_A^2 / n_A)}$
   $= (2.44)\sqrt{(0.10^2/36) + (0.11^2/36)}$
   $= (2.44)(0.025) = 0.06$

c. $(\overline{x}_B - \overline{x}_A) \pm E = 0.60 \pm 0.06$

<u>0.54 to 0.66</u>, the 0.98 confidence interval for $\mu_B - \mu_A$

Information on the MINITAB command to find a confidence interval for the difference between two means can be found in JES2-p439&440. The one command calculates the results for a confidence interval and hypothesis test, along with the degrees of freedom used.

10.39 Sample Statistics:

A: n = 10, $\overline{x}$ = 6.0, s^2 = 1.333
B: n = 10, $\overline{x}$ = 4.0, s^2 = 2.667

Step 1: The difference between two means, $\mu_A - \mu_B$
Step 2: a. normality assumed
   b. t   c. $1 - \alpha = 0.98$
Step 3: a. sample information given above
   b. $\overline{x}_A - \overline{x}_B = 6.0 - 4.0 = 2.0$
Step 4: a. $\alpha/2 = 0.02/2 = 0.01$; df = 9;
   $t(9, 0.01) = 2.82$

b. $E = t(df, \alpha/2) \cdot \sqrt{(s_A^2 / n_A) + (s_B^2 / n_B)}$
   $= (2.82)\sqrt{(1.333/10) + (2.667/10)}$
   $= (2.82)(0.632) = 1.78$

c. $(\overline{x}_A - \overline{x}_B) \pm E = 2.0 \pm 1.78$

<u>0.22 to 3.78</u>, the 0.98 confidence interval for $\mu_A - \mu_B$

10.41 Step 1: The difference between the mean scores for organ donors and nonorgan donors, $\mu_{non} - \mu_{donor}$
Step 2: a. normality assumed
   b. t   c. $1 - \alpha = 0.95$
Step 3: a. sample information given in exercise
   b. $\overline{x}_{non} - \overline{x}_{donor} = 7.62 - 5.36 = 2.26$

Step 4: a. $\alpha/2 = 0.05/2 = 0.025$; df = 24;
t(24, 0.025) = 2.06

b. $E = t(df, \alpha/2) \cdot \sqrt{(s_{non}^2 / n_{non}) + (s_{donor}^2 / n_{donor})}$

$= (2.06)\sqrt{(3.45^2/69) + (2.91^2/25)}$

$= (2.06)(0.715) = 1.473$

c. $(\overline{x}_{non} - \overline{x}_{donor}) \pm E = 2.26 \pm 1.473$

__0.787 to 3.733__, the 0.95 confidence interval for $\mu_{non} - \mu_{donor}$

10.43   Step 1: The difference between the average lengths of
workweek for Mining and Manufacturing, $\mu_{Min} - \mu_{Man}$
Step 2: a. normality indicated
b. t   c. $1-\alpha = 0.95$
Step 3: a. sample information given in exercise
b. $\overline{x}_{Min} - \overline{x}_{Man} = 47.5 - 43.5 = 4.0$
Step 4: a. $\alpha/2 = 0.05/2 = 0.025$; df = 9;
t(9, 0.025) = 2.26

b. $E = t(df, \alpha/2) \cdot \sqrt{(s_{Min}^2 / n_{Min}) + (s_{Man}^2 / n_{Man})}$

$= (2.26)\sqrt{(5.5^2/15) + (4.9^2/10)}$

$= (2.26)(2.1018) = 4.75$

c. $(\overline{x}_{Min} - \overline{x}_{Man}) \pm E = 4.0 \pm 4.75$

__-0.75 to 8.75__, the 0.95 confidence interval for $\mu_{Min} - \mu_{Man}$

WRITING HYPOTHESES FOR THE DIFFERENCE BETWEEN TWO MEANS

null hypothesis:

$H_O: \mu_1 = \mu_2$ __or__ $H_O: \mu_1 - \mu_2 = 0$ __or__ $H_O: \mu_1 - \mu_2 = \#$

possible alternative hypotheses:

$H_a: \mu_1 > \mu_2$ __or__ $H_a: \mu_1 - \mu_2 > 0$ __or__ $H_a: \mu_1 - \mu_2 > \#$

$H_a: \mu_1 < \mu_2$ __or__ $H_a: \mu_1 - \mu_2 < 0$ __or__ $H_a: \mu_1 - \mu_2 < \#$

$H_a: \mu_1 \neq \mu_2$ __or__ $H_a: \mu_1 - \mu_2 \neq 0$ __or__ $H_a: \mu_1 - \mu_2 \neq \#$

10.45   a. $H_O: \mu_1 - \mu_2 = 0$ vs. $H_a: \mu_1 - \mu_2 \neq 0$

b. $H_O: \mu_1 - \mu_2 = 0 \ (\leq)$ vs. $H_a: \mu_1 - \mu_2 > 0$

c. H_O: $\mu_1 - \mu_2 = 20$ ($\leq$) vs. H_a: $\mu_1 - \mu_2 > 20$

d. H_O: $\mu_A - \mu_B = 50$ ($\geq$) vs. H_a: $\mu_A - \mu_B < 50$
   or equivalently
   H_O: $\mu_B - \mu_A = -50$ ($\leq$) vs. H_a: $\mu_B - \mu_A > -50$

Review the rules for calculating the p-value in: JES2-p341, ST-p187, if necessary. Remember to use the t-distribution, therefore either Table 6, Table 7 or the computer will be used to find probabilities.
Review of the use of the tables can be found in: JES2-pp383-385, ST-pp213.

10.47   Table 6, Appendix B, JES2-p524
         Table 7, Appendix B, JES2-p525

a. $\mathbf{P} = P(t^* > 1.3 | df = 5)$;
      Using Table 6: $0.10 < \mathbf{P} < 0.25$
      Using Table 7: $\mathbf{P} = 0.125$

b. $\mathbf{P} = P(t^* < -2.8 | df = 8) = P(t^* > 2.8 | df = 8)$;
      Using Table 6: $0.01 < \mathbf{P} < 0.025$
      Using Table 7: $\mathbf{P} = 0.012$

c. $\mathbf{P} = 2P(t^* > 1.8 | df = 15)$;
      Using Table 6: $0.05 < \mathbf{P} < 0.10$
      Using Table 7: $\mathbf{P} = 2(0.046) = 0.092$

d. $\mathbf{P} = 2P(t^* > 1.8 | df = 25)$;
      Using Table 6: $0.05 < \mathbf{P} < 0.10$
      Using Table 7: $\mathbf{P} = 2(0.042) = 0.084$

Hypothesis Test for the Difference Between Two Means,
Independent Samples

Review the parts to a hypothesis test as outlined in: JES2-pp334&350, ST-pp183&192, if needed. Slight changes will occur in:

   1. **the hypotheses**: (see box before exercise 10.45)

                                                    . . .

2. **the calculated test statistic**

$$t = \frac{(\bar{x}_1 - \bar{x}_2) - (\mu_1 - \mu_2)}{\sqrt{(s_1^2 / n_1) + (s_2^2 / n_2)}} \quad , \text{ using df = smaller of df}_1 \text{ or df}_2$$

3. If H_O is rejected, a significant difference between the means is indicated.
   If H_O is not rejected, no significant difference between the means is indicated.

Any subscripts may be used on the hypotheses. Try to use letters that indicate the source. The form H_O: $\mu_2 - \mu_1 = 0$ (versus H_O: $\mu_1 = \mu_2$) is the preferred form since it establishes the order for subtraction that will be needed when calculating the test statistic.

Hint for writing the hypotheses for exercise 10.49

Look at the last sentence of the exercise; "Do the data show that the mean score for those with computer experience was significantly less than the mean score for those without computer experience?" Assuming that the mean anxiety score for those without computer experience is larger than the mean anxiety score for those with computer experience, subtracting in the form of "mean score-no exp" - "mean score - exp," should yield a positive (> 0) answer. The words "less than" indicates <, and therefore belongs in the alternative hypotheses. The negation "not less than" indicates greater than or equal to ($\geq$), which goes with the null hypothesis. You still write the null hypothesis with the equal sign (=), but include the greater than or equal-to sign ($\geq$) in parentheses after it.

10.49 Step 1: The difference between mean scores on an anxiety test for statistics students taking the course with and without the use of computers, $\mu_2 - \mu_1$

Step 2: H_O: $\mu_2 - \mu_1 = 0$
H_a: $\mu_2 - \mu_1 > 0$ ($\mu_1 < \mu_2$)

Step 3: a. normality assumed
b. t   c. $\alpha = 0.05$

Step 4: a. sample information given in exercise

b. $t* = [(\bar{x}_2 - \bar{x}_1) - (\mu_2 - \mu_1)] / \sqrt{(s_2^2 / n_2) + (s_1^2 / n_1)}$

   $= [(67.2 - 60.3) - 0] / [\sqrt{(2.1^2/15) + (7.5^2/10)}]$

   $= 2.84$

c. $P = P(t* > 2.84 | df = 9)$;

   Using Table 6, Appendix B, JES2-p524:
   $0.005 < P < 0.01$
   Using Table 7, Appendix B, JES2-p525:
   $0.008 < P < 0.012$

Step 5: a. $P < \alpha$ b. Reject H_O

c. At the 0.05 level of significance, there is sufficient evidence that the students with computer experience scored less than those without.

10.51 a. p-value approach:

Step 1: The difference in mean weights of 10-lb bags of potatoes, $\mu_b - \mu_s$

Step 2: H_O: $\mu_b - \mu_s = 0$

        H_a: $\mu_b - \mu_s \neq 0$

Step 3: a. normality assumed, CLT with $n_b = 100$ and $n_s = 100$.

b. t   c. $\alpha = 0.05$

Step 4: a. sample information given in exercise

b. $t = [(\bar{x}_b - \bar{x}_s) - (\mu_b - \mu_s)] / \sqrt{(s_b^2 / n_b) + (s_s^2 / n_s)}$

   $t* = [(10.4 - 10.2) - 0] / [\sqrt{(0.25/100) + (0.36/100)}]$

   $= 2.56$

c. $P = 2 \cdot P(t(df = 99) > 2.56)$;

   Using Table 6, Appendix B, JES2-p524:
   $0.01 < P < 0.02$
   Using Table 7, Appendix B, JES2-p525:
   $2[0.006+ < \tfrac{1}{2}P < 0.008]$; $0.012 < P < 0.016$

Step 5: a. $P < \alpha$ b. Reject H_O

c. There is sufficient evidence to reject the null hypothesis; that is, the two population means are significantly different.

Use Table 6 (Appendix B, JES2-p524) with the smaller of df_1 or df_2 and the given α to find the critical value(s). Reviewing how to determine the test criteria in: JES2-p386, ST-p193, as it is applied to the t-distribution may be helpful.

b. Classical approach:

Step 1: The difference in mean weights of 10-lb bags of potatoes, $\mu_b - \mu_s$

Step 2: H_o: $\mu_b - \mu_s = 0$

H_a: $\mu_b - \mu_s \neq 0$

Step 3: a. normality assumed, CLT with $n_b = 100$ and $n_s = 100$.

b. t   c. $\alpha = 0.05$

d. $\pm t(90, 0.025) = \pm 1.99$

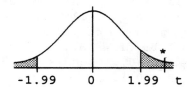

-1.99   0   1.99   t

Step 4: a. sample information given in exercise

b. $t = [(\bar{x}_b - \bar{x}_s) - (\mu_b - \mu_s)] / \sqrt{(s_b^2 / n_b) + (s_s^2 / n_s)}$

$t^* = [(10.4 - 10.2) - 0] / [\sqrt{(0.25/100) + (0.36/100)}]$

$= 2.56$

Step 5: a. t^* is in the critical region, see * step 3d

b. Reject H_o

c. There is sufficient evidence to reject the null hypothesis; that is, the two population means are significantly different.

10.53   a. $t^* = [(\bar{x}_{act} - \bar{x}_{pl}) - (\mu_{act} - \mu_{pl})] / st.err.$

$= 33.00/9.68 = \underline{3.41}$

b. Bounds using minimum df values:

$P = 2P(t^* > 3.41 | df = 7)$;

Using Table 6, Appendix B, JES2-p524:

$0.01 < P < 0.02$

Using Table 7, Appendix B, JES2-p525:

$2(0.005 < \tfrac{1}{2}P < 0.006)$, $0.010 < P < 0.012$

Bounds using maximum df values:

$P = 2P(t^* > 3.41 | df = 18)$;

Using Table 6, Appendix B, JES2-p524:

$P < 0.01$

Using Table 7, Appendix B, JES2-p525:

$2(0.001 < \frac{1}{2}P < 0.002)$, $0.002 < P < 0.004$

Therefore; <u>$0.002 < P < 0.012$</u>

c. $t^* = [(\overline{x}_{act} - \overline{x}_{pl}) - (\mu_{act} - \mu_{pl})]/st.err.$

  $= 28.75/13.74 = \underline{2.09}$

d. Bounds using minimum df values:

$P = 2P(t^* > 2.09 | df = 7)$;

Using Table 6, Appendix B, JES2-p524:

$0.05 < P < 0.10$

Using Table 7, Appendix B, JES2-p525:

$2(0.037 < \frac{1}{2}P < 0.043)$, $0.074 < P < 0.086$

Bounds using maximum df values:

$P = 2P(t^* > 2.09 | df = 18)$;

Using Table 6, Appendix B, JES2-p524:

$0.05 < P < 0.10$

Using Table 7, Appendix B, JES2-p525:

$2(0.025 < \frac{1}{2}P < 0.030)$, $0.050 < P < 0.060$

Therefore; <u>$0.050 < P < 0.086$</u>

10.55 a. Step 1: The difference between the mean particle size
for two types of disc centrifuges, $\mu_{JLDC} - \mu_{DPJ}$

Step 2: H_O: $\mu_{JLDC} - \mu_{DPJ} = 0$

H_a: $\mu_{JLDC} - \mu_{DPJ} \neq 0$

Step 3: a. normality indicated

b. t   c. $\alpha = 0.10$

Step 4: a. $n_{JLDC} = 13$, $\overline{x}_{JLDC} = 4684$, $s^2_{JLDC} = 124,732$

$n_{DPJ} = 13$, $\overline{x}_{DPJ} = 4408.9$, $s^2_{DPJ} = 112,019.91$

b. $t^* = [(\overline{x}_{JLDC} - \overline{x}_{DPJ}) - (\mu_{JLDC} - \mu_{DPJ})]/\sqrt{(s^2_{JLDC}/n_{JLDC}) + (s^2_{DPJ}/n_{DPJ})}$

$= [(4684 - 4408.9) - 0]/[\sqrt{(124732/13) + (112019.91/13)}]$

$= 2.04$

c. $P = 2P(t^* > 2.04 | df = 12)$;

Using Table 6, Appendix B, JES2-p524:

$0.05 < P < 0.10$

Using Table 7, Appendix B, JES2-p525:

$2[0.029 < \frac{1}{2}P < 0.034]$; $0.058 < P < 0.068$

Step 5: a. $P < \alpha$ b. Reject H_O
        c. At the 0.10 level of significance, there is
           sufficient evidence to show a difference
           between the readings.

b. Step 1: The difference between mean particle size for
           two types of disc centrifuges, $\mu_{JLDC} - \mu_{DPJ}$
   Step 2: a. normality indicated
           b. t   c. $1 - \alpha = 0.90$
   Step 3: a. sample information given above
           b. $\overline{x}_{JLDC} - \overline{x}_{DPJ} = 4684 - 4408.9 = 275.1$
   Step 4: a. $\alpha/2 = 0.10/2 = 0.05$; df = 12;
              $t(12, 0.05) = 1.78$
           b. $E = t(df, \alpha/2) \cdot \sqrt{(s^2_{JLDC} / n_{JLDC}) + (s^2_{DPJ} / n_{DPJ})}$
              $= (1.78)\sqrt{(124732/13) + (112019.91/13)}$
              $= (1.78)(134.95) = 240.211$
           c. $(\overline{x}_{JLDC} - \overline{x}_{DPJ}) \pm E = 275.1 \pm 240.211$
   <u>34.89 to 515.31</u>, the 0.90 confidence interval for $\mu_{JLDC} - \mu_{DPJ}$

10.57 Everybody will get different results, but they can all
be expected to look very similar to the following.
   a.   N(100,20)

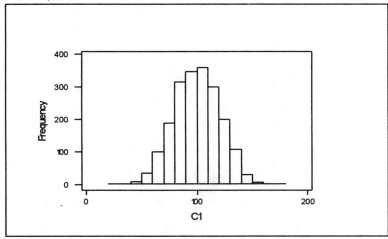

N(120,20)

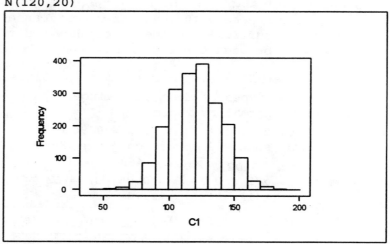

b.   The sampling distribution is expected to be normal
       in shape with a mean of 20 (120-100) and have a

       standard error of $\sqrt{\dfrac{20^2}{8} + \dfrac{20^2}{8}}$ or 10.

d. 100 values for the difference between two sample
   means:

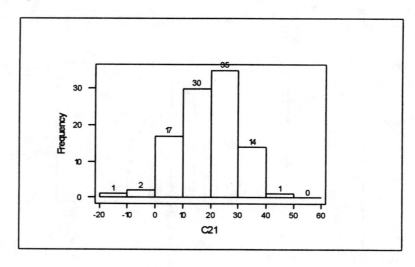

e. For the empirical sampling distribution, the mean
    is 19.51 and the standard error is 10.71. There is
    65%, 96% and 99% of the values within one, two and
    three standard errors of the expected mean of 20,
    respectively. This seems to agree closely with the
    empirical rule, thus suggesting a normal
    distribution occurred.

f. You can expect very similar results to occur on
    repeated trials.

10.59 Everybody will get different results, but they can all be
expected to look very similar to the results found in
exercise 10.63. It turns out that the t* statistic is very
"robust", meaning "it works quite well even when the
assumptions are not met." This it one of the reasons the
t-test for the mean and the t-test for the difference
between two means are such important tests.

SECTION 10.4 MARGIN EXERCISES

10.61 $x = \underline{75}$, $n = \underline{250}$,
$p' = x/n = 75/250 = \underline{0.30}$, $q' = 1 - p' = 1 - 0.30 = \underline{0.70}$

10.62 a. $n_1p_1 = 40(0.9) = \underline{36}$, $n_1q_1 = 40(0.1) = \underline{4}$
    $n_2p_2 = 50(0.9) = \underline{45}$, $n_2q_2 = 50(0.1) = \underline{5}$

b. No, n_1q_1 and n_2q_2 are not larger than 5.

10.63 Step 1: The difference between proportions, $p_A - p_B$
Step 2: a. n's > 20, np's and nq's all > 5
          b. z   c. $1 - \alpha = 0.95$
Step 3: a. sample information given in exercise
            $p'_A = x_A / n_A = 45/125 = 0.36$, $q'_A = 1 - 0.36 = 0.64$
            $p'_B = x_B / n_B = 48/150 = 0.32$, $q'_B = 1 - 0.32 = 0.68$
          b. $p'_A - p'_B = 0.36 - 0.32 = 0.04$

Step 4: a. $\alpha/2 = 0.05/2 = 0.025$; $z(0.025) = 1.96$

b. $E = z(\alpha/2) \cdot \sqrt{(p_A q_A)/n_A + (p_B q_B)/n_B}$

$= (1.96)\sqrt{(0.36)(0.64)/125 + (0.32)(0.68)/150}$

$= (1.96)(0.057) = 0.11$

c. $(p_A' - p_B') \pm E = 0.04 \pm 0.11$

<u>-0.07 to 0.15</u>, the 0.95 confidence interval for p_A-p_B

10.64 $p_P' = (x_E + x_R)/(n_E + n_R) = (15 + 25)/(250 + 275)$

$= 40/525 = \underline{0.076}$

$q_P' = 1 - p_P' = 1 - 0.076 = \underline{0.924}$

10.65 Rewrite the alternative hypothesis for easier understanding:

H_a: $p_R - p_E > 0$

$p_R' = 25/275 = 0.091$, $p_E' = 15/250 = 0.06$

$z^* = (p_R' - p_E')/\sqrt{(p_P')(q_P')[(1/n_R)+(1/n_E)]}$

$= (0.091 - 0.06)/\sqrt{(0.076)(0.924)[(1/275)+(1/250)]}$

$= 0.031/0.0232 = 1.34$

P $= P(z > 1.34) = (0.5000 - 0.4099) = \underline{0.0901}$

10.66 Step 1: The difference in proportions of extra time needed in the recovery room for smokers and nonsmokers, $p_S - p_n$

Step 2: H_O: $p_S - p_n = 0$

H_a: $p_S - p_n > 0$

Step 3: a. n's > 20,   np's and nq's > 5

b. z

Step 4: a. $n_S = 163$, $p_s' = 0.38$, $n_n = 164$, $p_n' = 0.23$

$x_S = (163)(0.38) = 62$

$x_n = (164)(0.23) = 38$

$p_P' = (x_S + x_n)/(n_S + n_n) = (62 + 38)/(163 + 164)$

$= 100/327 = 0.306$

$q_P' = 1 - p_P' = 1 - 0.306 = 0.694$

b. $z^* = (p_s' - p_n')/\sqrt{(p_P')(q_P')[(1/n_s) + (1/n_n)]}$

$z^* = (0.38 - 0.23)/\sqrt{(0.306)(0.694)[(1/163)+(1/164)]}$

$= 0.15/0.0510 = 2.94$

c. $P = P(z > 2.94)$;

Using Table 3, Appendix B, JES2-p521:

$P = (0.5000 - 0.4984) = 0.0016$

Using Table 5, Appendix B, JES2-p523:

$0.0016 < P < 0.0019$

Step 5: Yes, P is well below the usual values of α. Nonsmokers spend significantly less time in the recovery room.

10.67 $p_G' = 323/380 = 0.85$, $p_H' = 332/420 = 0.79$

$p_P' = (x_G + x_H)/(n_G + n_H) = (323 + 332)/(380 + 420)$

$= 655/800 = 0.82$

$q_P' = 1 - p_P' = 1 - 0.82 = 0.18$

$z^* = (p_G' - p_H')/\sqrt{(p_P')(q_P')[(1/n_G)+(1/n_H)]}$

$= (0.85 - 0.79)/\sqrt{(0.82)(0.18)[(1/380)+(1/420)]}$

$= 0.06/0.0272 = \underline{2.21}$

SECTION 10.4 EXERCISES

Estimating $(p_1 - p_2)$ - the difference between two population proportions - independent samples
(large samples)

1. Point estimate: $p_1' - p_2'$

2. Confidence interval:

$$(p_1' - p_2') \quad \pm \quad z(\alpha/2) \cdot \sqrt{(p_1' \cdot q_1' / n_1) + (p_2' \cdot q_2' / n_2)}$$

    ↑   ↑   ↑

   point   confidence   estimated
  estimate   coefficient   standard error

Maximum error of the estimate

10.69 Step 1: The difference in proportions of nurses who experienced a change in position based on their participation in a program, $p_w - p_n$

Step 2: a. n's > 20, np's and nq's all > 5

b. z   c. $1-\alpha = 0.99$

Step 3: a. $n_w = 341$, $x_w = 87$, $p_w' = 87/341 = 0.255$,

$q_w' = 1 - 0.255 = 0.745$

$n_n = 40$, $x_n = 9$, $p_n' = 9/40 = 0.225$,

$q_n' = 1 - 0.225 = 0.775$

b. $p_w' - p_n' = 0.255 - 0.225 = 0.03$

Step 4: a. $\alpha/2 = 0.01/2 = 0.005$; $z(0.005) = 2.58$

b. $E = z(\alpha/2) \cdot \sqrt{(p_w' \cdot q_w' / n_w) + (p_n' \cdot q_n' / n_n)}$

$= 2.58\sqrt{(0.255)(0.745)/341 + (0.225)(0.775)/40}$

$= (2.58)(0.07) = 0.18$

c. $(p_w' - p_n') \pm E = 0.03 \pm 0.18$

<u>-0.15 to 0.21</u>, the 0.99 interval for $p_w - p_n$

10.71 Step 1: The difference in proportions of defectives for two
machines, $p_1 - p_2$

Step 2: a. n's > 20, np's and nq's all > 5

b. z   c. $1-\alpha = 0.90$

Step 3: a. $n_1 = 150$, $x_1 = 12$, $p_1' = 12/150 = 0.08$,

$q_1' = 1 - 0.08 = 0.92$

$n_2 = 150$, $x_2 = 6$, $p_2' = 6/150 = 0.04$,

$q_2' = 1 - 0.04 = 0.96$

b. $p_1' - p_2' = 0.08 - 0.04 = 0.04$

Step 4: a. $\alpha/2 = 0.10/2 = 0.05$; $z(0.05) = 1.65$

b. $E = z(\alpha/2) \cdot \sqrt{(p_1' \cdot q_1' / n_1) + (p_2' \cdot q_2' / n_2)}$

$= 1.65\sqrt{(0.08 \cdot 0.92/150) + (0.04 \cdot 0.96/150)}$

$= (1.65)(0.024) = 0.04$

c. $(p_1' - p_2') \pm E = 0.04 \pm 0.04$

<u>0.000 to 0.080</u>, the 0.90 interval for $p_1 - p_2$

WRITING HYPOTHESES FOR THE DIFFERENCE BETWEEN TWO PROPORTIONS
a) null hypothesis:
$$H_O: p_1 = p_2 \quad \underline{or} \quad p_1 - p_2 = 0$$
b) possible alternative hypotheses:
$$H_a: p_1 > p_2 \quad \underline{or} \quad H_a: p_1 - p_2 > 0$$
$$H_a: p_1 < p_2 \quad \underline{or} \quad H_a: p_1 - p_2 < 0$$
$$H_a: p_1 \neq p_2 \quad \underline{or} \quad H_a: p_1 - p_2 \neq 0$$

10.73 a. $H_O: p_m - p_w = 0 \quad$ vs. $\quad H_a: p_m - p_w \neq 0$

b. $H_O: p_b - p_g = 0 \ (\leq) \quad$ vs. $\quad H_a: p_b - p_g > 0$

c. $H_O: p_c - p_{nc} = 0 \ (\leq) \quad$ vs. $\quad H_a: p_c - p_{nc} > 0$

Hypothesis Test for the Difference Between Two Proportions,
Independent Samples (Large Samples)

Review parts to a hypothesis test as outlined in: JES2-pp334&350, ST-pp183&192, if needed. Changes will occur in:

1. **the hypotheses**: (see box before exercise 10.73)

2. **the calculated test statistic**
$$z = \frac{(p_1' - p_2') - (p_1 - p_2)}{\sqrt{p_p' q_p'(\frac{1}{n_1} + \frac{1}{n_2})}} \quad , \text{ where } p_1' = \frac{x_1}{n_1} \ , \ p_2' = \frac{x_2}{n_2} \ ,$$

$$p_p' = \frac{x_1 + x_2}{n_1 + n_2} \quad \text{and} \quad q_p' = 1 - p_p'$$

3. If H_O is rejected, a significant difference in proportions is indicated.
   If H_O is not rejected, no significant difference is indicated.

...

10.75 Step 1: The difference in the rate of impotence reported by non-smokers and former smokers, $p_n - p_s$

Step 2: H_O: $p_n - p_s = 0$

H_a: $p_n - p_s \neq 0$

Step 3: a. n's > 20, np's and nq's all > 5

b. z

Step 4: a. $n_n = 1162$, $x_n = 26$, $p_n' = 26/1162 = 0.022$

$n_s = 1292$, $x_s = 26$, $p_s' = 26/1292 = 0.02$

$p_P' = (x_n + x_s)/(n_n + n_s)$

$\quad = (26+26)/(1162+1292) = 0.021$

$q_P' = 1 - p_P' = 1.000 - 0.021 = 0.979$

b. $z = [(p_n' - p_s') - (p_n - p_s)]/\sqrt{(p_P')(q_P')[(1/n_n) + (1/n_s)]}$

$z* = (0.022 - 0.02)/\sqrt{(0.021)(0.979)[(1/1162)+(1/1292)]}$

$\quad = \underline{0.35}$

c. **P** = 2P(z* > 0.35);

Using Table 3, Appendix B, JES2-p521:

P = 2(0.5000 - 0.1368) = 2(0.3632) = $\underline{0.7264}$

Using Table 5, Appendix B, JES2-p523:

P = 2(0.3632) = $\underline{0.7264}$

<u>Hint for writing the hypotheses for exercise 10.77</u>

Look at the last sentence of the exercise; "Is there sufficient evidence to show a <u>difference</u> in the effectiveness of the <u>2 image campaigns</u>...?" The words "2 image campaigns" imply 2 populations, namely the citizens exposed to a conservative campaign and citizens exposed to a moderate campaign. The results are given in the form of proportions, therefore a difference between 2 proportions is suggested. The word "difference" indicates a <u>not equal to</u>. Therefore, the alternative is <u>not equal to</u> ($\neq$). The negation becomes "equal to" and the null hypothesis would be written with an equality sign (=).

10.77   a. p-value approach:

Step 1: The difference in the proportions for the effectiveness of two campaign images, $p_m - p_C$

Step 2: H_O: $p_m - p_C = 0$

  H_a: $p_m - p_C \neq 0$

Step 3: a. n's > 20, np's and nq's all > 5

  b. z   c. $\alpha = 0.05$

Step 4: a. $n_m = 100$, $p'_m = 0.50$, $n_C = 100$, $p'_c = 0.40$

  $p'_p = (x_m + x_C)/(n_m + n_C) = (50+40)/(100+100) = 0.45$

  $q'_p = 1 - p'_p = 1.000 - 0.45 = 0.55$

  b. $z = [(p'_m - p'_c) - (p_m - p_C)]/\sqrt{(p'_p)(q'_p)[(1/n_m) + (1/n_c)]}$

  $z* = (0.50 - 0.40)/\sqrt{(0.45)(0.55)[(1/100) + (1/100)]}$

  $= 0.10/0.0704 = 1.42$

  c. $P = 2P(z > 1.42)$;

  Using Table 3, Appendix B, JES2-p521:
  $P = 2(0.5000 - 0.4222) = 2(0.0778) = 0.1556$
  Using Table 5, Appendix B, JES2-p523:
  $2(0.0735 < \frac{1}{2}P < 0.0806)$; $0.1470 < P < 0.1612$

Step 5: a. $P > \alpha$ b. Fail to reject H_O

  c. There is not sufficient evidence to show a difference, at the 0.05 level.

Use Table 4 (Appendix B, JES2-p522), (Normal Distribution) to determine the test criteria. If the amount of α is not listed in Table 4, use Table 3 (Appendix B, JES2-p521). A review of determining the test criteria in: JES2-pp354&355, ST-p193, may be helpful.

---

b. Classical approach:

Step 1: The difference in the proportions for the effectiveness of two campaign images, $p_m - p_C$

Step 2: H_O: $p_m - p_C = 0$

  H_a: $p_m - p_C \neq 0$

Step 3: a. n's > 20, np's and nq's all > 5

  b. z   c. $\alpha = 0.05$

d.

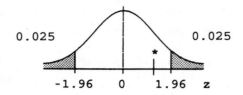

$$0.025 \qquad 0.025$$
$$-1.96 \quad 0 \quad 1.96 \quad z$$

Step 4: a. $n_m = 100$, $p'_m = 0.50$, $n_c = 100$, $p'_c = 0.40$

$$p'_P = (x_m + x_c)/(n_m + n_c) = (50+40)/(100+100) = 0.45$$
$$q'_P = 1 - p'_P = 1.000 - 0.45 = 0.55$$

b. $z = [(p'_m - p'_c) - (p_m - p_c)]/\sqrt{(p'_P)(q'_P)[(1/n_m) + (1/n_c)]}$

$$z* = (0.50 - 0.40)/\sqrt{(0.45)(0.55)[(1/100) + (1/100)]}$$
$$= 0.10/0.0704 = 1.42$$

Step 5: a. z* falls in the noncritical region, see * Step 3d
   b. Fail to reject H_O
   c. There is not sufficient evidence to show a
      difference, at the 0.05 level.

10.79 a. H_O: $p_m - p_w = 0$ $\qquad$ H_a: $p_m - p_w > 0$

$$p'_P = (x_m + x_w)/(n_m + n_w) = (215 + 170)/(500 + 500) = 0.385$$
$$q'_P = 1 - p'_P = 1.000 - 0.385 = 0.615$$

$z = [(p'_m - p'_w) - (p_m - p_w)]/\sqrt{(p'_P)(q'_P)[(1/n_m) + (1/n_w)]}$

$$z* = (0.43 - 0.34)/\sqrt{(0.385)(0.615)[(1/500) + (1/500)]}$$
$$= 0.09/0.0308 = 2.92$$

P = P(z > 2.92);
   Using Table 3, Appendix B, JES2-p521:
   P = 0.5000 - 0.4982 = <u>0.0018</u>
   Using Table 5, Appendix B, JES2-p523:
   <u>0.0016 < P < .0019</u>

b. Reject H_O. The smoking rate for male diabetics is
   significantly higher than for female diabetics, at the
   0.05 level.

CHAPTER EXERCISES

10.81 Step 1: The mean difference in IQ scores for oldest and youngest members of a family (d = O - Y)

Step 2: a. normality indicated

b. t   c. $1-\alpha = 0.95$

Step 3: a. $n = 12$, $\bar{d} = 3.583$, $s_d = 19.58$

b. $\bar{d} = 3.583$

Step 4: a. $\alpha/2 = 0.05/2 = 0.025$; df = 11;

$t(11, 0.025) = 2.20$

b. $E = t(df,\alpha/2)\cdot(s_d/\sqrt{n}) = (2.20)(19.58/\sqrt{12})$

$= (2.20)(5.65) = 12.435$

c. $\bar{d} \pm E = 3.583 \pm 12.435$

<u>-8.85 to 16.02</u>, the 0.95 interval for μ

10.83 Step 1: The mean difference in scores for recruits participating in a rifle-shooting competition (d = week later - first day)

Step 2: H_0: $\mu_d = 0$

H_a: $\mu_d > 0$ (improvement)

Step 3: a. normality assumed

b. t   c. $\alpha = 0.05$

Step 4: a. $n = 10$, $\bar{d} = 5.2$, $s_d = 7.406$

b. $t* = (\bar{d} - \mu_d)/(s_d/\sqrt{n})$

$= (5.2 - 0.0)/(7.406/\sqrt{10}) = 2.22$

c. $\mathbf{P} = P(t > 2.22|df=9)$;

Using Table 6, Appendix B, JES2-p524:

$0.025 < \mathbf{P} < 0.05$

Using Table 7, Appendix B, JES2-p525:

$0.022 < \mathbf{P} < 0.029$

Step 5: a. $\mathbf{P} < \alpha$ b. Reject H_0

c. There is a significant improvement, at the 0.05 level.

10.85 Step 1: The difference between the mean anxiety scores for males and females, $\mu_f - \mu_m$

Step 2: a. normality assumed, CLT with $n_f = 50$ and $n_m = 50$.

b. t   c. $1-\alpha = 0.95$

Step 3: a. $n_f = 50$, $\bar{x}_f = 75.7$, $s_f = 13.6$

$n_m = 50$, $\bar{x}_m = 70.5$, $s_m = 13.2$

b. $\bar{x}_f - \bar{x}_m = 75.7 - 70.5 = 5.2$

Step 4: a. $\alpha/2 = 0.05/2 = 0.025$; df = 49;
$\qquad$ t(49, 0.025) = 2.02

$\qquad$ b. $E = t(df, \alpha/2) \cdot \sqrt{(s_f^2 / n_f) + (s_m^2 / n_m)}$

$\qquad\qquad$ = $(2.02)\sqrt{(13.6^2/50) + (13.2^2/50)}$

$\qquad\qquad$ = $(2.02)(2.68) = 5.41$

$\qquad$ c. $(\overline{x}_f - \overline{x}_m) \pm E = 5.2 \pm 5.41$

$\qquad\qquad$ <u>-0.21 to 10.61</u>, the 0.95 interval for $\mu_f - \mu_m$

10.87 Step 1: The difference between the mean scores for a group
$\qquad\qquad$ that had a high school computer course and those
$\qquad\qquad$ that did not, $\mu_1 - \mu_2$

$\qquad$ Step 2: a. normality indicated

$\qquad\qquad$ b. t $\qquad\qquad$ c. $1 - \alpha = 0.98$

$\qquad$ Step 3: a. $n_1 = 15$, $\overline{x}_1 = 25.73$, $s_1^2 = 37.64$

$\qquad\qquad$ $n_2 = 22$, $\overline{x}_2 = 23.05$, $s_2 2 = 24.33$

$\qquad\qquad$ b. $\overline{x}_1 - \overline{x}_2 = 25.73 - 23.05 = 2.68$

$\qquad$ Step 4: a. $\alpha/2 = 0.02/2 = 0.01$; df = 14;
$\qquad\qquad$ t(14, 0.01) = 2.62

$\qquad\qquad$ b. $E = t(df, \alpha/2) \cdot \sqrt{(s_1^2 / n_1) + (s_2^2 / n_2)}$

$\qquad\qquad\qquad$ = $(2.62)\sqrt{(37.64/15) + (24.33/22)}$

$\qquad\qquad\qquad$ = $(2.62)(1.90137) = 4.98$

$\qquad\qquad$ c. $(\overline{x}_1 - \overline{x}_2) \pm E = 2.68 \pm 4.98$

$\qquad\qquad$ <u>-2.3 to 7.66</u>, the 0.98 confidence interval for $\mu_1 - \mu_2$

10.89 a.

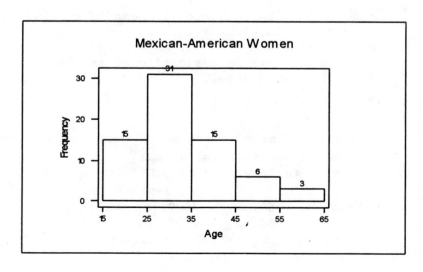

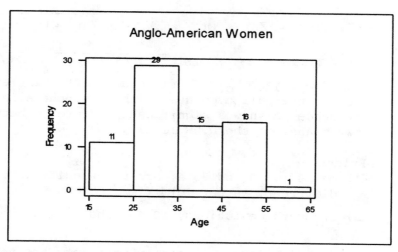

Anglo-American Women

b. Summary of extensions table:
$$n = \Sigma f = 70, \quad \Sigma xf = 2310, \quad \Sigma x^2 f = 83{,}700$$

$$\bar{x} = \Sigma xf / \Sigma f = 2310/70 = \underline{33.0}$$

$$s^2 = [\Sigma x^2 f - ((\Sigma xf)^2 / \Sigma f)] / (\Sigma f - 1)$$
$$= [83{,}700 - (2310^2/70)]/69 = 7470/69 = 108.261$$

$$s = \sqrt{s^2} = \sqrt{108.261} = \underline{10.405}$$

c. Summary of extensions table:
$$n = \Sigma f = 72, \quad \Sigma xf = 2550, \quad \Sigma x^2 f = 98100$$

$$\bar{x} = \Sigma xf / \Sigma f = 2550/72 = \underline{35.42}$$

$$s^2 = [\Sigma x^2 f - ((\Sigma xf)^2 / \Sigma f)] / (\Sigma f - 1)$$
$$= [98100 - (2550^2/72)]/71 = 7787.5/71 = 109.683$$

$$s = \sqrt{s^2} = \sqrt{109.683} = \underline{10.473}$$

d. $H_0: \mu_a - \mu_m = 0$ vs. $H_a: \mu_a - \mu_m \neq 0$

$$\alpha = 0.05$$

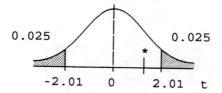

0.025   0.025

-2.01   0   2.01 t

$$t = [(\overline{x}_a - \overline{x}_m) - (\mu_a - \mu_m)] / \sqrt{(s_a^2 / n_a) + (s_m^2 / n_m)}$$

$$t* = [(35.4 - 33.0)-0] / [\sqrt{(109.68/72) + (108.26/70)}] = \underline{1.37}$$

Fail to reject H_O. There is not sufficient evidence to reject the null hypothesis; that is, there is no evidence to show a significant difference between the two means, at the 0.05 level.

    e. Evidence:
       1) means, 33.0 and 35.4, are not significantly different, as per (d)

       2) standard deviations, 10.405 and 10.473, nearly identical

       3) distributions, as seen in part (a), are very similar in appearance and shape

Conclusion: there appears to be no evidence to suggest that there is any major differences between the two distributions. That is, the difference observed could be due to randomness.

10.91 Less accurate would be implied if the mean target error is greater.

    Step 1: The difference between the mean target errors for two short-range rockets, $\mu_2 - \mu_1$
    Step 2: H_O: $\mu_2 - \mu_1 = 0$
            H_a: $\mu_2 - \mu_1 > 0$ (2nd type less accurate)
    Step 3: a. normality indicated
           b. t   c. $\alpha = 0.01$
    Step 4: a. $n_1 = 8$, $\overline{x}_1 = 36$, $s_1 = 15$
                $n_2 = 10$, $\overline{x}_2 = 52$, $s_2 = 18$
           b. $t* = [(\overline{x}_2 - \overline{x}_1) - (\mu_2 - \mu_1)] / \sqrt{(s_2^2 / n_2) + (s_1^2 / n_1)}$
               $= [(52 - 36) - 0] / [\sqrt{(18^2/10) + (15^2/8)}]$
               $= 2.06$
           c. $\mathbf{P} = P(t > 2.06 | df = 7)$;
              Using Table 6, Appendix B, JES2-p524:
              $0.025 < \mathbf{P} < 0.05$
              Using Table 7, Appendix B, JES2-p525:
              $0.037 < \mathbf{P} < 0.043$

Step 5: a. $P > \alpha$ b. Fail to reject H_0
c. At the 0.01 level of significance, there is not sufficient evidence to show that the second kind of rocket is less accurate.

10.93 a. Step 1: The difference between mean exercise capacity for women with no period and women with a normal cycle, $\mu_{nc} - \mu_{np}$
Step 2: H_0: $\mu_{nc} - \mu_{np} = 0$
H_a: $\mu_{nc} - \mu_{np} \neq 0$
Step 3: a. normality assumed
b. t   c. $\alpha = 0.05$
Step 4: a. sample information given in exercise
b. $t* = [(\overline{x}_{nc} - \overline{x}_{np}) - (\mu_{nc} - \mu_{np})] / \sqrt{(s_{nc}^2 / n_{nc}) + (s_{np}^2 / n_{np})}$
$= [(31.3 - 30.5) - 0] / [\sqrt{(8.0^2/8) + (9.5^2/8)}]$
$= 0.18$
c. $P = 2P(t > 0.18 | df = 7)$;
Using Table 6, Appendix B, JES2-p524:
$P > 0.50$
Using Table 7, Appendix B, JES2-p525:
$2(0.424 < \tfrac{1}{2}P < 0.462)$, $0.848 < P < 0.924$
b. Step 5: a. $P > \alpha$ b. Fail to reject H_0
c. At the 0.05 level of significance, there is not a significant difference between the two groups.

10.95 a. It appears that the Task method showed significant results for both strokes.
b. Using Table 6 and a two-tailed test, the forehand results are significant at 0.05 and the backhand is significant at 0.01.

10.97 a. Step 1: The difference in proportions of illicit drug users at the University of Michigan, $p_8 - p_9$
Step 2: a. n's > 20,   np's and nq's all > 5
b. z   c. $1 - \alpha = 0.95$
Step 3: a. $n_8 = 17,000$, $p_8' = 0.213$,
$q_8' = 1 - 0.213 = 0.787$
$n_9 = 17,000$, $p_9' = 0.197$,
$q_9' = 1 - 0.197 = 0.803$

b. $p_8' - p_9' = 0.213 - 0.197 = 0.016$

Step 4: a. $\alpha/2 = 0.05/2 = 0.025$; $z(0.025) = 1.96$

b. $E = z(\alpha/2) \cdot \sqrt{(p_8' \cdot q_8' / n_8) + (p_9' \cdot q_9' / n_9)}$

$= 1.96 \cdot \sqrt{(0.213 \cdot 0.787 / 17000) + (0.197 \cdot 0.803 / 17000)}$

$= (1.96)(0.00438) = 0.009$

c. $(p_8' - p_9') \pm E = 0.016 \pm 0.009$

<u>0.007 to 0.025</u>, the 0.95 interval for $p_8 - p_9$

b. Yes, there is a significant difference at the 0.05 level because 0 is not within the interval.

10.99 Step 1: The difference in proportions requiring service from two manufacturers, $p_1 - p_2$

Step 2: a. n's > 20, np's and nq's all > 5

b. z   c. $1 - \alpha = 0.95$

Step 3: a. sample information given in exercise

b. $p_1' - p_2' = 0.15 - 0.09 = 0.060$

Step 4: a. $\alpha/2 = 0.05/2 = 0.025$; $z(0.025) = 1.96$

b. $E = z(\alpha/2) \cdot \sqrt{(p_1' \cdot q_1' / n_1) + (p_2' \cdot q_2' / n_2)}$

$= 1.96 \cdot \sqrt{(0.15 \cdot 0.85/75) + (0.09 \cdot 0.91/75)}$

$= (1.96)(0.0528) = 0.104$

c. $(p_1' - p_2') \pm E = 0.060 \pm 0.104$

<u>-0.044 to 0.164</u>, the 0.95 interval for $p_1 - p_2$

10.101 Step 1: The difference in proportions of left-handers and right-handers that die at the wheel, $p_1 - p_r$

Step 2: H_O: $p_1 - p_r = 0$

H_a: $p_1 - p_r > 0$

Step 3: a. n's > 20, np's and nq's all > 5 (marginally)

b. z

Step 4: a. $n_1 = 100$, $x_1 = 5$, $p_1' = 5/100 = 0.05$

$n_r = 900$, $x_r = 18$, $p_r' = 18/900 = 0.02$

$p_p' = (x_1 + x_r)/(n_1 + n_r) = (5+18)/(100+900) = 0.023$

$q_p' = 1 - p_p' = 1.000 - 0.023 = 0.977$

b. $z = [(p_1' - p_r') - (p_1 - p_r)]/\sqrt{(p_p')(q_p')[(1 / n_1) + (1 / n_r)]}$

$z* = (0.05 - 0.02)/\sqrt{(0.023)(0.977)[(1/100) + (1/900)]}$

$= 1.90$

c. $P = P(z > 1.90)$;
        Using Table 3, Appendix B, JES2-p521:
        $P = 0.5000 - 0.4713 = 0.0287$
        Using Table 5, Appendix B, JES2-p523:
        $P = 0.0287$

The sample results are significant for all values of α greater than 0.03. That is, when $\alpha \geq 0.03$, the null hypothesis will be rejected and the evidence considered significant.

10.103 Step 1: The difference in proportions of men and women who have some gray hair by age 25, $p_w - p_m$

Step 2: $H_O: p_w - p_m = 0$
        $H_a: p_w - p_m > 0$

Step 3: a. n's > 20, np's and nq's all > 5
        b. z   c. $\alpha = 0.01$

Step 4: a. $n_w = 1000$, $p'_w = 0.34$, so $x_w = 340$
        $n_m = 1000$, $p'_m = 0.29$, so $x_m = 290$

$$p'_P = (x_w + x_m)/(n_w + n_m) = (340 + 290)/(1000 + 1000) = 0.315$$

$$q'_P = 1 - p'_P = 1.000 - 0.315 = 0.685$$

b. $z = [(p'_w - p'_m) - (p_w - p_m)]/\sqrt{(p'_P)(q'_P)[(1/n_w) + (1/n_m)]}$

$z^* = (0.34 - 0.29)/\sqrt{(0.315)(0.685)[(1/1000) + (1/1000)]}$
     $= 0.05/0.0208 = 2.40$

c. $P = P(z > 2.40)$;
        Using Table 3, Appendix B, JES2-p521:
        $P = 0.5000 - 0.4918 = 0.0082$
        Using Table 5, Appendix B, JES2-p523:
        $P = 0.0082$

Step 5: a. $P < \alpha$ b. Reject H_O
        c. At the 0.01 level of significance, the 5% difference is significant.

CHAPTER 11 ∇ APPLICATIONS OF CHI-SQUARE

Chapter Preview

Chapter 11 demonstrates hypothesis tests, as did Chapters 8, 9, and 10. The difference lies in the type of data that is to be analyzed. Enumerative type data, that is, data which can be counted and placed into categories, will be discussed and investigated in three types of tests. Each test will compare actual (observed) results with expected (theoretical) results. One will use the comparisons to determine whether a "claimed" relationship exists, one will determine whether two factors or variables are independent, and the third will determine whether the proportions per variable are the same. Also, a new test statistics, the chi square statistic, χ^2, will be introduced and utilized in performing these tests.

SECTION 11.1 MARGIN EXERCISES

11.1   a. $\chi^2 (10, 0.01) = \underline{23.2}$ b. $\chi^2 (12, 0.025) = \underline{23.3}$

11.2   a. $\chi^2 (10, 0.95) = \underline{3.94}$ b. $\chi^2 (22, 0.995) = \underline{8.64}$

SECTION 11.1 EXERCISES

χ^2 Distribution
(used for inferences concerning σ and σ^2)

Key facts about the χ^2 curve:
1) the total area under a χ^2 curve is 1
2) it is skewed to the right (stretched out to the right side, not symmetrical)
3) it is nonnegative, starts at zero and continues out towards $+\infty$
4) a different curve exists for each sample size
5) uses α and degrees of freedom, df, to determine table values
6) degrees of freedom is abbreviated as 'df', where, df = n-1
7) for df > 2, the mean of the distribution is df.

Notation: $\chi^2(df, \alpha) = \chi^2$("degrees of freedom","area to the right")

                    ↑   ↑   ↑
                Table 8   row id #   column id #

ex.) Right tail: $\chi^2(13, .025) = 24.7$ (n must have been 14)

     Left tail: $\chi^2(13, .975) = 5.01$

Note: For left tail, "area to right" includes both the area in the "middle" and the area of the "right" tail.

11.3 a. 34.8   b. 28.8   c. 13.4   d. 48.3

       e. 12.3   f. 3.25   g. 37.7   h. 10.9

SECTION 11.2 MARGIN EXERCISES

11.5 9 parts + 3 parts + 3 parts + 1 part = 16 parts total; thus: 9/16, 3/16, 3/16 and 1/16.

11.6 $E = n \cdot p = 556(9/16) = \underline{312.75}$

       $O - E = 315 - 312.75 = \underline{2.25}$

       $(O-E)^2/E = (2.25)^2/312.75 = \underline{0.0162}$

11.7   a. Trial: each adult surveyed
        b. Variable: Reason why we rearrange furniture
        c. Multiple answers: Bored, moving, new furniture, etc.

11.8   Since the percentages sum to more than 100%, some people use more than one method. In order to be a multinomial experiment, each source most yield exactly one response.

Characteristics of a Multinomial Experiment

1. There are **n** identical independent trials.

2. The outcome of each trial fits into exactly one of the **k** possible categories or cells.

3. The number of times a trial outcome falls into a particular cell is given by O_i (O - for observed, i for $1 \rightarrow k$).

4. $O_1 + O_2 + O_3 + \ldots + O_k = n$

5. There is a constant probability associated with each of the k cells in such a way that $p_1 + p_2 + \ldots + p_k = 1$.

NOTE: Variable - the characteristic about each item that is of interest.

Various levels of the variable - the k possible outcomes or responses.

Writing Hypotheses for Multinomial Experiments

The null hypothesis is written in a form to show that there is no difference between the experimental (observed) frequencies and the theoretical (expected) frequencies.
The alternative hypothesis is the "opposite" of the null hypothesis. It is written in a form to show that a difference does exist.

ex.: The marital status distribution for New York state is 21%, 64%, 8%, and 7% for the possible categories of single, married, widowed, and divorced.
   H_O: $P(S) = .21$, $P(M) = .64$, $P(W) = .08$, $P(D) = .07$
   H_a: The percentages are different than specified in H_O.

ex.: A gambler thinks that a die may be *loaded*.
   H_O: $P(1) = P(2) = P(3) = P(4) = P(5) = P(6) = 1/6$ (not loaded)
   H_a: The probabilities are different.   (is loaded)

11.9   a. H_O: $P(1) = P(2) = P(3) = P(4) = P(5) = 0.2$
      H_a: The numbers are not equally likely.

b. H_O: $P(1) = 2/8$, $P(2) = 3/8$, $P(3) = 2/8$, $P(4) = 1/8$
   H_a: The probabilities are distributed differently than listed in H_O.

c. H_O: $P(E) = 0.16$, $P(G) = 0.38$, $P(F) = 0.41$, $P(P) = 0.05$
   H_a: The percentages are different than specified in H_O.

Steps for a Hypothesis/Significance Test for Multinomial Experiments

Follow the steps outlined for p-value and classical hypothesis tests in: JES2-pp334&350, ST-pp183&192.
The variations are noted below.

1. Distribute the sample information (observed frequencies) into the appropriate cells (data may be already categorized).

2. Calculate the expected frequencies using probabilities determined by H_O and the formula:
   $E_i = np_i$, where E_i is the expected frequency for cell i,
   p_i is the probability for the ith cell

3. Use the observed and expected frequencies from each cell to calculate the test statistic, χ^2.

   Use the formula $\chi^2 = \sum_{allcells} \frac{(O - E)^2}{E}$

4.a) p-value approach:
   Since all of the α is placed in the right tail,
   the p-value = $P(\chi^2 > \chi^{2*})$.
   The p-value is <u>estimated</u> using Table 8 (Appendix B, JES2-p526):
   a) Locate df row.
   b) Locate χ^{2*} between two critical values in the df row; the p-value is in the interval between the two corresponding probabilities at the top of the columns labeled *area to the right*.

   OR

   The p-value can be calculated using MINITAB commands found in JES2-p473.

   . . .

b) Classical approach:
Follow the steps in determining the test criteria found in:
JES2-p474, ST-p277. Then locate χ^{2*} on the χ^2 curve with respect to the critical value.

5. Make a decision and interpret it.
   a) If $P < \alpha$ or the calculated test statistic falls into the critical region, then reject H_O. There is sufficient evidence to indicate that there is a *difference* between the observed and expected frequencies.
   b) If $P > \alpha$ or the calculated test statistic falls into the noncritical region, fail to reject H_O. There is <u>not</u> sufficient evidence to indicate a *difference* between the observed and expected frequencies.

11.11 a. H_O: $P(A) = P(B) = P(C) = P(D) = P(E) = 0.2$

b. χ^2

c. (1) & (2)
Step 1: Preference of floor polish, the probability that a particular type is preferred.
Step 2: H_O: Equal preference
         H_a: preferences not all equal
Step 3: a. Assume that the 100 consumers represent a random sample.
         b. χ^2 with df = 4   c. $\alpha = 0.10$
Step 4: a. $\chi^2 = \Sigma[(O-E)^2/E]$ (as found on accompanying table)

$$E = n \cdot p = 100(1/5) = 20, \text{ for all cells}$$

Polish	A	B	C	D	E	Total
Observed	27	17	15	22	19	100
Expected	20	20	20	20	20	100
$(O-E)^2/E$	49/20	9/20	25/20	4/20	1/20	88/20

$\chi^{2*} = 4.40$

----

(1) Step 4b. $P = P(\chi^2 > 4.40 | df=4)$;
              Using Table 8: $0.25 < P < 0.50$
              Using computer: $P = 0.355$
     Step 5a. $P > \alpha$

```
┌─────────────────────────────────────────────────────────────────────┐
│                   Determining the Test Criteria                     │
│                                                                     │
│ 1. Draw a picture of the χ² distribution (skewed right,starting at 0).│
│                                                                     │
│ 2. Locate the critical region (based on α and $H_a$).               │
│       Since we are testing $H_o$: "no difference" versus            │
│       $H_a$: "difference", all of the α is placed in the right tail │
│       to represent a significant or large *difference* between the  │
│       observed and expected values.                                 │
│                                                                     │
│ 3. Shade in the critical region (the area where you will reject $H_o$,│
│       the right-hand tail)                                          │
│                                                                     │
│ 4. Find the appropriate critical value from Table 8 (Appendix B,    │
│    JES2-p526), using χ²(df,α), where df = k - 1.  k is equal to the │
│    number of cells or categories the data are classified into.      │
│                                                                     │
│ Remember this critical or boundary value divides the area under the │
│ χ² distribution curve into critical and noncritical regions and is  │
│ part of the critical region.                                        │
└─────────────────────────────────────────────────────────────────────┘
```

        OR
        (2)   Step 3d.

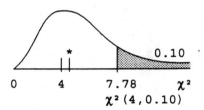

$$\chi^2(4,0.10)$$

        Step 5a. χ²* falls in the noncritical region, see *
            Step 3d.

Step 5: b. Fail to reject H_o
          c. There is insufficient evidence to conclude there
            is a difference in polish preference.

A small program was written for MINITAB to calculate expected values and chi-square values. The percentages or decimals stated in the null hypothesis should be entered into a column (C1), whereas the corresponding observed frequencies are entered into another column (C2).

LET commands can be used to calculate the expected frequencies (C3) and the corresponding chi-square values (C4).

Checks were programmed in: sum of probability column (C5) is equal to 1; sum of the observed frequency column (C6) is equal to the sum of the expected frequency column (C7).
The sum of the individual chi-square values becomes the calculated chi-square value (C8).
See JES2-p473 for information on MINITAB commands to find the p-value.

11.13   a. MINITAB verify -- answers given in exercise
$$P(E) = 0.16, \; P(G) = 0.38, \; P(F) = .41, \; P(P) = 0.05$$

$$\chi^2 = \Sigma[(O-E)^2/E] \text{ (as found on accompanying table)}$$

$$E(E) = n \cdot P(E) = 350(0.16) = \underline{56}$$
$$E(G) = n \cdot P(G) = 350(0.38) = \underline{133}$$
$$E(F) = n \cdot P(F) = 350(0.41) = \underline{143.5}$$
$$E(P) = n \cdot P(P) = 350(0.05) = \underline{17.5}$$

Role Model	excell.	good	fair	poor	Total
Observed	44	145	133	28	350
Expected	56	133	143.5	17.5	350
$(O-E)^2/E$	2.5714	1.0827	0.7683	6.3	10.722

$$\chi^{2*} = \underline{10.722}$$

b. df = k - 1 = 4 - 1 = $\underline{3}$

c. **P** = $P(\chi^2 > 10.722 | df=3)$; $\underline{0.010 < P < 0.025}$

11.15 Step 1: The proportions: P(studio/efficiency),
P(1 bedroom), P(2 bedrooms), P(3 bedrooms),
P(over 3 bedrooms)

Step 2: H_o: The size of vacation homes preferred by baby
boomers are in the proportions of 18.2%,
18.2%, 40.4%, 18.2%, 5%

H_a: The proportions are different than listed

Step 3: a. Assume that the 300 individuals surveyed
represent a random sample.

b. χ^2 with df = 4   c. $\alpha = 0.05$

Step 4: a. $\chi^2 = \Sigma[(O-E)^2/E]$ (as found on accompanying table)

E(S|E) = n·p = 300(0.182) = 54.6, E(1B) = 300(0.182) = 54.6
E(2B) = 300(0.404) = 121.2, E(3B) = 300(0.182) = 54.6,
E(>3B) = 300(0.05) = 15

Vac.Home	S|E	1B	2B	3B	>3B	Total
Observed	75	60	105	45	15	300
Expected	54.6	54.6	121.2	54.6	15	300
$(O-E)^2/E$	7.622	0.534	2.165	1.688	0	12.01

$\chi^{2*} = 12.01$

Step 4b. **P** = $P(\chi^2 > 12.01 | df=4)$;
Using Table 8: $0.01 < $ **P** $ < 0.025$
Using computer: **P** $= 0.017$

Step 5a. **P** $< \alpha$

OR

Step 3c. $\chi^2(4, 0.05) = 9.49$
Step 5a. χ^{2*} falls in the critical region

Step 5: b. Reject H_o

c. The distribution of preferences is different in
Nebraska from that found nationally.

One of the properties of random numbers is that they occur with equal probabilities. Since there are 10 values (0-9), each integer will have the same probability 1/10 = 0.1.

11.17 Step 1: The probability of a single-digit integer being generated.

Step 2: H_O: $P(0) = P(1) = P(2) = \ldots = P(9) = 0.1$
H_a: The probabilities are not all equal

Step 3: a. Assume that the 100 numbers generated represent a random sample.
b. χ^2 with df = 9   c. $\alpha = 0.05$

Step 4: a. $\chi^2 = \Sigma[(O-E)^2/E]$ (as found on accompanying table)

$E = n \cdot p = 100(0.1) = 10$, for all cells

Integer	0	1	2	3	4	5	6	7	8	9	Total
Observed	11	8	7	7	10	10	8	11	14	14	100
Expected	10	10	10	10	10	10	10	10	10	10	100
$(O-E)^2/E$	.1	.4	.9	.9	.0	.0	.4	.1	1.6	1.6	6.0

$\chi^2* = 6.00$

Step 4b. **P** = $P(\chi^2 > 6.00 | df=9)$;
Using Table 8: $0.50 < $ **P** $ < 0.75$
Using computer: **P** $ = 0.740$

Step 5a. **P** $> \alpha$

OR

Step 3c. $\chi^2(9, 0.05) = 16.9$
Step 5a. χ^2* falls in the noncritical region

Step 5: b. Fail to reject H_O
c. The sample values do not show sufficient reason to reject the uniform hypothesis, at the 0.05 level of significance.

SECTION 11.3 MARGIN EXERCISES

11.19 $E = (40)(50)/200 = 2000/200 = \underline{10}$

11.20 a. <u>113</u>   b. <u>122</u>   c. <u>300</u>   d. <u>68.23</u>

11.21

a.
Reason	Men	Women
No time	0.36	0.37
In good health	0.39	0.21
Poor health	0.11	0.15
Dislike exercise	0.04	0.08
Too lazy	0.01	0.06
Other	0.09	0.13
Totals	1.00	1.00

b.
Reason	Men	Women	Totals
No time	360	370	730
In good health	390	210	600
Poor health	110	150	260
Dislike exercise	40	80	120
Too lazy	10	60	70
Other	90	130	220
Totals	1000	1000	2000

c. Step 1: The proportions of reasons why men do not
exercise and the proportions of reasons why
women do not exercise.

Step 2: H_O: The reasons for not exercising are
distributed the same for both men and
women.

H_{a}: The reasons for not exercising are
distributed differently for men and women.

Step 3: a. χ^2 with df = 5   b. $\alpha = 0.05$

Step 4: a. $\chi^2 = \Sigma[(O-E)^2/E]$ (as found on table below)

Reason	Expected values Men	Women	Totals	Chi-square
No time	365	365	730	0.0685 + 0.0685
In good	300	300	600	27.000 + 27.000
Poor he	130	130	260	3.0769 + 3.0769
Dislike	60	60	120	6.6667 + 6.6667
Too laz	35	35	70	17.857 + 17.857
Other	110	110	220	3.6364 + 3.6364
Totals	1000	1000	2000	$\chi^{2}* = 116.611$

```
-------------------------------------------------------------
        Step 4b.  P = P(χ² > 116.611|df=5);
                    Using Table 8:     P < 0.005
                    Using computer:    P = 0+
        Step 5a.  P < α
    OR
        Step 3c.  χ²(5,0.05) = 11.1
        Step 5a.  χ²* falls in the critical region
-------------------------------------------------------------
```

Step 5: b. Reject H_O

        c. The reasons for not exercising are significantly different for men and women.

SECTION 11.3 EXERCISES

Contingency Tables

A contingency table is a table consisting of rows and columns used to summarize and cross-classify data according to two variables. Each row represents the categories for one of the variables, and each column represents the categories for the other variable. The intersections of these rows and columns produce cells. The data will be in a form where two varieties of hypothesis tests are possible. These are:

1. Tests of independence
   - to determine if one variable is independent of the other variable, and

2. Tests of homogeneity
   - to determine if the proportion distribution for one of the variables is the same for each of the categories of the second variable.

11.23 The *test of independence* has one sample of data that is being cross-tabulated according to the categories of two separate variables; the *test of homogeneity* has multiple samples being compared side-by-side and together these samples form the entire sample used in the contingency table.

Writing Hypotheses for Tests of Independence and/or Homogeneity

The null hypothesis is written in a form to show that there is no difference between the experimental (observed) frequencies and the theoretical (expected) frequencies. The alternative hypothesis is the opposite of the null hypothesis. It is written in a form to show that a difference does exist.

Therefore, in tests of independence:

H_O: One variable is independent of the other variable

and in tests of homogeneity:

H_O: The proportions for one variable are distributed the same for all categories of the second variable.

MINITAB can perform a hypothesis test for independence or homogeneity. Since these tests are completed in the same fashion, the same command may be used. See JES2-p492 for more information.

11.25 Step 1: The proportions of irritations:
P(eye for age groups), P(nose for age groups),
P(throat for age groups).

Step 2: H_O: The type of ENT irritation is independent of the age group.
H_a: The type of ENT irritation is not independent of the age group.

Step 3: a. χ^2 with df = 6   b. $\alpha = 0.05$

Step 4: a. $\chi^2 = \Sigma[(O-E)^2/E]$ (as found on accompanying table)

Expected counts are printed below observed counts

	18-29	30-44	45-64	65-over	Total
1	440	567	349	59	1415
	432.85	585.97	347.99	48.18	
2	924	1311	794	102	3131
	957.78	1296.59	770.02	106.62	
3	253	311	157	19	740
	226.37	306.44	181.99	25.20	
Total	1617	2189	1300	180	5286

ChiSq = 0.118 + 0.614 + 0.003 + 2.430 + 1.191 + 0.160 +
        0.747 + 0.200 + 3.133 + 0.068 + 3.432 + 1.525
      = 13.621

$$\chi^2* = 13.621$$

Step 4b. $P = P(\chi^2 > 13.621 | df=6)$;
         Using Table 8: $0.025 < P < 0.05$
         Using computer: $P = 0.035$
Step 5a. $P < \alpha$

OR

Step 3c. $\chi^2(6,0.05) = 12.6$
Step 5a. χ^2* falls in the critical region

Step 5: b. Reject H_O
        c. The type of ENT irritation is not independent of
           the age group at the 0.05 level of significance.

11.27 a. & b.
Step 1: The proportions: P(yes for three different
        reactions),P(no for three different reactions).
Step 2: H_O: The distribution of reactions is the same for
           both groups.
        H_a: The distribution of reactions is not the same
           for both groups.
Step 3: a. χ^2 with df = 2   b. $\alpha = 0.10$
Step 4: a. $\chi^2 = \Sigma[(O-E)^2/E]$ (as found on accompanying table)

Expected counts are printed below observed counts

	Mild	Medium	Strong	Total
1	170	100	30	300
	144.00	120.00	36.00	
2	70	100	30	200
	96.00	80.00	24.00	
Total	240	200	60	500

ChiSq = 4.694 + 3.333 + 1.000 +
        7.042 + 5.000 + 1.500 = 22.569

$$\chi^2* = 22.569$$

a. Step 4b. $P = P(\chi^2 > 22.569 | df=2)$;

Using Table 8: $P < 0.005$

Using computer: $P = 0+$

Step 5a. $P < \alpha$

OR

b. Step 3c. $\chi^2(2, 0.10) = 4.61$

Step 5a. χ^2* falls in the critical region

Step 5: b. Reject H_O

c. Yes, there is sufficient evidence to show a relationship between neighborhood and reaction.

11.29 Step 1: The proportions of thefts: P(autos for regions), P(truck/buses for regions), P(other vehicles for regions).

Step 2: H_O: Thefts of various vehicle types is proportioned the same across regions.

H_a: Thefts of various vehicle types is not proportioned the same across regions.

Step 3: a. χ^2 with df = 6   b. $\alpha = 0.05$

Step 4: a. $\chi^2 = \Sigma[(O-E)^2/E]$ (as found on accompanying table)

Expected counts are printed below observed counts

	Auto	Trk/Bus	Other	Total
1	85	10	5	100
	72.50	20.00	7.50	
2	80	15	5	100
	72.50	20.00	7.50	
3	65	25	10	100
	72.50	20.00	7.50	
4	60	30	10	100
	72.50	20.00	7.50	
Total	290	80	30	400

ChiSq = 2.155 + 5.000 + 0.833 + 0.776 + 1.250 + 0.833 + 0.776 + 1.250 + 0.833 + 2.155 + 5.000 + 0.833

= 21.695

$\chi^2* = 21.695$

Step 4b. $P = P(\chi^2 > 21.695 \mid df=6)$;

Using Table 8: $P < 0.005$

Using computer: $P = 0.001$

Step 5a. $P < \alpha$

OR

Step 3c. $\chi^2(6, 0.05) = 12.6$

Step 5a. χ^{2*} falls in the critical region

Step 5: b. Reject H_O

c. The distribution of types of vehicles thefts is not proportioned the same across regions at the 0.05 level of significance.

11.31 Information to be collected from doctors:

1. Does the doctor deliver babies?
2. How old is the doctor?
3. How many babies has the doctor delivered by Cesarean? How many by natural child birth?

To table the information, a contingency table would be used; classes of doctor's age forming the columns, and method of birth forming the rows. The entries would be number of babies in each cross category. Chi-square would be used to analyze the resulting counts.

CHAPTER EXERCISES

11.33 a. & b.

Step 1: The proportions: P(1st type), P(2nd type), P(3rd type)

Step 2: H_O: 1:3:4 proportions

H_a: proportions are other than 1:3:4

Step 3: a. Assume that the 800 hybrids represent a random sample.

b. χ^2 with df = 2   c. $\alpha = 0.05$

Step 4: a. $\chi^2 = \Sigma[(O-E)^2/E]$ (as found on accompanying table)

$E(1st) = n \cdot p = 800[1/(1+3+4)] = 100$

$E(2nd) = 800[3/(1+3+4)] = 300$

$E(3rd) = 800[4/(1+3+4)] = 400$

	1st	2nd	3rd	Total
Observed	80	340	380	800
Expected	100	300	400	800
$(O-E)^2/E$	4.00	5.33	1.00	10.33

$$\chi^{2*} = 10.33$$

---

(1) Step 4b. $P = P(\chi^2 > 10.33 | df=2)$;

Using Table 8: $0.005 < P < 0.01$
Using computer: $P = 0.006$

Step 5a. $P < \alpha$

OR

(2) Step 3d. $\chi^2(2, 0.05) = 5.99$
Step 5a. χ^{2*} falls in the critical region

Step 5: b. Reject H_O
c. We have sufficient evidence to show that the ratio is not the hypothesized 1:3:4 ratio.

11.35 Step 1: The proportions of AIDS deaths in age groups: P(under 12), P(13-29), P(30-39), P(40-49), P(50-59), P(60 and over).

Step 2: H_O: P(under 12) = 0.01, P(13-29) = 0.18, P(30-39) = 0.45, P(40-49) = 0.24, P(50-59) = 0.08, P(60 and over) = 0.04

H_a: The percentages are different than listed

Step 3: a. Assume that the 1000 AIDS deaths represent a random sample.
b. χ^2 with df = 5   c. $\alpha = 0.05$

Step 4: a. $\chi^2 = \Sigma[(O-E)^2/E]$ (as found on accompanying table)

Row	P	OBS	EXP	CHI-SQ
1	0.01	8	10	0.4000
2	0.18	205	180	3.4722
3	0.45	390	450	8.0000
4	0.24	270	240	3.7500
5	0.08	120	80	20.0000
6	0.04	7	40	27.2250

Row	SUM(P)	SUN(OBS)	SUM(EXP)	CHI-SQ*
1	1.00	1000	1000	62.8472

```
--------------------------------------------------------------
         Step 4b.  P = P(χ² > 62.8472|df=5);
                     Using Table 8:     P < 0.005
                     Using computer:    P = 0+
         Step 5a.  P < α
   OR
         Step 3c.  χ²(5,0.05) = 11.1
         Step 5a.  χ²* falls in the critical region
--------------------------------------------------------------
```

Step 5: b. Reject H_O
c. We have sufficient evidence to show that the distribution of AIDS deaths is not the same as over the time period 1982-1992.

11.37 Step 1: The proportions of age and HPV-Positive:
P(≤20), P(21-25), P(26-30), P(31-35).
Step 2: H_O: The same proportion of each age group is HPV-Positive.
H_a: The proportions are different per age group.
Step 3: a. χ^2 with df = 3   b. α = 0.05
Step 4: a. $\chi^2 = \Sigma[(O-E)^2/E]$ (as found on accompanying table)

Expected counts are printed below observed counts

	HPV-Posi	HPV-Neg	Total
1	11	16	27
	8.66	18.34	
2	30	51	81
	25.98	55.02	
3	34	74	108
	34.63	73.37	
4	18	56	74
	23.73	50.27	
Total	93	197	290

ChiSq = 0.632 + 0.299 + 0.622 + 0.294 +
         0.011 + 0.005 + 1.384 + 0.653 = 3.9

```
--------------------------------------------------------------
         Step 4b.  P = P(χ² > 3.9|df=3);
                     Using Table 8:     0.25 < P < 0.50
                     Using computer:    P = 0.273
         Step 5a.  P > α
   OR
```

Step 3c. $\chi^2(3,0.05) = 7.82$
Step 5a. χ^{2*} falls in the noncritical region

Step 5: b. Fail to reject H_O
c. The distribution of proportions of HPV-Positive is the same for each age group.

11.39 a. & b.
Step 1: The proportions of popcorn that popped and did not pop: P(Brand A), P(Brand B), P(Brand C), P(Brand D)
Step 2: H_O: Proportion of popcorn that popped is the same for all brands.
H_a: The proportions are not the same for all brands.
Step 3: a. χ^2 with df = 3   b. $\alpha = 0.05$
Step 4: a. $\chi^2 = \Sigma[(O-E)^2/E]$

Expected values:

	A	B	C	D	Total
Popped	88	88	88	88	352
Not popped	12	12	12	12	48
Totals	100	100	100	100	400

$\chi^{2*} = \Sigma[(O-E)^2/E] = 0.333 + 1.333 + 0.083 + 0.750$
$+ 0.045 + 0.182 + 0.011 + 0.102$
$\chi^{2*} = 2.839$

----

a.   Step 4b. $P = P(\chi^2 > 2.839 | df=3)$;
Using Table 8: $0.25 < P < 0.50$
Using computer: $P = 0.417$
Step 5a. $P > \alpha$
OR
b.   Step 3c. $\chi^2(3,0.05) = 7.82$
Step 5a. χ^{2*} falls in the noncritical region

Step 5: b. Fail to reject H_O
c. There is insufficient evidence to show that the proportions of popped corn are not the same for all brands, at the 0.05 level of significance.

11.41 Step 1: The proportions: P(yes for different school locations),P(no for different school locations), P(unsure for different school locations).

Step 2: H_O: The response and the school's location are independent.

H_a: The response and the school's location are not independent.

Step 3: a. χ^2 with df = 4   b. $\alpha = 0.05$

Step 4: a. $\chi^2 = \Sigma[(O-E)^2/E]$

Frequencies [expected frequencies]

	Urban	Suburban	Rural	Totals
Yes	57 [57.86]	27 [31.48]	47 [41.66]	131
No	23 [22.53]	16 [12.25]	12 [16.22]	51
Unsure	45 [44.61]	25 [24.27]	31 [32.12]	101
Totals	125	68	90	283

$$\chi^{2}* = \Sigma[(O - E)^2/E] = 0.013 + 0.638 + 0.684 +$$
$$0.010 + 1.148 + 1.098 +$$
$$0.003 + 0.022 + 0.039 = 3.655$$

Step 4b. **P** = $P(\chi^2 > 3.655 | df=4)$;

Using Table 8: $0.25 <$ **P** < 0.50

Using computer: **P** = 0.456

Step 5a. **P** $> \alpha$

OR

Step 3c. $\chi^2(4,0.05) = 9.49$

Step 5a. $\chi^{2}*$ falls in the noncritical region

Step 5: b. Fail to reject H_O

c. The evidence is not sufficient to show a significant relationship between student response and school's location.

11.43 a. & b.

Step 1: The proportions: P(days absent for different categories of employees), P(days worked for different categories of employees).

Step 2: H_O: Rate of absenteeism is the same for all groups.

H_a: Rate of absenteeism is not the same for all groups.

Step 3: a. χ^2 with df = 3   b. $\alpha = 0.01$

Step 4: a. $\chi^2 = \Sigma[(O-E)^2/E]$

Expected values:

Days	MM	SM	MF	SF
Absent	200	70	80	150
Worked	9400	3290	3760	7050

$\chi^{2*} = 27.2369$

---

a.   Step 4b. $P = P(\chi^2 > 27.2369 | df=3)$;
　　　　　Using Table 8:　　$P < 0.005$
　　　　　Using computer:　$P = 0+$
　Step 5a. $P < \alpha$
OR
b.　Step 3c. $\chi^2(3, 0.01) = 11.3$
　Step 5a. χ^{2*} falls in the critical region

Step 5: b. Reject H_O
　　　　　c. The rate of absenteeism is not the same for all categories of employees.

11.45 Conditions for rolling a balanced die 600 times:
The critical value is $\chi^2(5, 0.05) = 11.1$ (6 possible outcomes)

With 600 rolls, the expected value for each cell is 100.

Many combinations of observed frequencies are possible to cause us to reject the equally likely hypothesis. The combinations will have to have a calculated χ^2 value greater than 11.1 or a p-value less than 0.05. Two possibilities are presented.

1. If each observed frequency is different from the expected by the same amount, then $11.1/6 = 1.85$ is the amount of chi-square that would come from each cell.

$(O-E)^2/E = (O - 100)^2/100 = 1.85$

$(O - 100)^2 = 185$

$O - 100 = \pm 13.6$

$O = 86$ or 114

That is, if three of the observed frequency values are 86 and the other three are 114, the faces of the die will be declared not to be equally likely.

Row	P	OBS	EXP	CHI-SQ
1	0.166667	86	100.000	1.96005
2	0.166667	86	100.000	1.96005
3	0.166667	86	100.000	1.96005
4	0.166667	114	100.000	1.95994
5	0.166667	114	100.000	1.95994
6	0.166667	114	100.000	1.95994

Row	SUM(P)	SUN(OBS)	SUM(EXP)	CHI-SQ*
1	1.00000	600	600.001	11.7600

DF   5

p-value   0.038

Note χ^2 value and p-value

2. Now suppose just one is different and the other five all occur with the same frequency:

Remember, the total observed most be 600. Therefore, for every five one outcome is different from the expected, the other five each must be different by one to balance. If the five are each different from the expected by x, then the one that is very different is off by 5x. The sum of 5 - x's squared and 5x squared is $30x^2$. Thus,

$$30x^2 = 11.1$$

$$x^2 = 11.1/30 = 0.37$$

$$(O-E)^2/E = (O-100)^2/100 = 0.37$$

$$(O-100)^2 = 37$$

$$O - 100 = \pm6.08 \quad \text{(round-up)}$$

O = either 93 or 107   for the five cells, and

O for the other cell must be off by 5(7) or 35; it is either 65 or 135.

Row	P	OBS	EXP	CHI-SQ
1	0.166667	93	100.000	0.4900
2	0.166667	93	100.000	0.4900
3	0.166667	93	100.000	0.4900
4	0.166667	93	100.000	0.4900
5	0.166667	93	100.000	0.4900
6	0.166667	135	100.000	12.2498

Row	SUM(P)	SUN(OBS)	SUM(EXP)	CHI-SQ*
1	1.00000	600	600.001	14.7000

DF   5

p-value   0.012

Note χ^2 and p-value.

INTRODUCTORY CONCEPTS

SUMMATION NOTATION

The Greek capital letter sigma $\left(\Sigma\right)$ is used in mathematics to indicate the summation of a set of addends. Each of these addends must be of the form of the variable following Σ. For example:

1. $\sum x$ means sum the variable x.
2. $\sum (x - 5)$ means sum the set of addends that are each 5 less than the values of each x.

When large quantities of data are collected, it is usually convenient to index the response variable so that at a future time its source will be known. This indexing is shown on the notation by using i (or j or k) and affixing the index of the first and last addend at the bottom and top of the Σ. For example,

$$\sum_{i=1}^{3} x_i$$

means to add all the consecutive values of x's starting with source number 1 and proceeding to source number 3.

▽ ILLUSTRATION 1

Consider the inventory in the following table concerning the number of defective stereo tapes per lot of 100.

Lot Number (I)	1	2	3	4	5	6	7	8	9	10
Number of Defective Tapes per Lot (x)	2	3	2	4	5	6	4	3	3	2

    a.   Find $\displaystyle\sum_{i=1}^{10} x_i$.   b.   Find $\displaystyle\sum_{i=4}^{8} x_i$.

Solution

a.
$$\sum_{i=1}^{10} x_i = x_1 + x_2 + x_3 + x_4 + \cdots + x_{10}$$

$$= 2 + 3 + 2 + 4 + 5 + 6 + 4 + 3 + 3 + 2 = 34$$

b.
$$\sum_{i=4}^{8} x_i = x_4 + x_5 + x_6 + x_7 + x_8 = 4 + 5 + 6 + 4 + 3 = 22$$

$$\Delta\Delta$$

The index system must be used whenever only part of the available information is to be used. In statistics, however, we will usually use all the available information, and to simplify the formulas we will make an adjustment. This adjustment is actually an agreement that allows us to do away with the index system in situations where all values are used.

Thus in our previous illustration, $\sum_{i=1}^{10} x_i$ could have been written simply as $\sum x$.

NOTE The lack of the index indicates that all data are being used.

∇ ILLUSTRATION A-2

Given the following six values for x, 1, 3, 7, 2, 4, 5, find $\sum x$.

Solution

$$\sum x = 1 + 3 + 7 + 2 + 4 + 5 = 22 \qquad \Delta\Delta$$

Throughout the study and use of statistics you will find many formulas that use the Σ symbol. Care must be taken so that the formulas are not misread. Symbols like $\sum x^2$ and $(\sum x)^2$ are quite different. $\sum x^2$ means "square each x value and then add up the squares," while $(\sum x)^2$ means "sum the x values and then square the sum."

Find (a) $\sum x^2$ and (b) $\left(\sum x\right)^2$ for the sample in Illustration A-2.

Solution

a.

x	1	3	7	2	4	5
x^2	1	9	49	4	16	25

$$\sum x^2 = 1 + 9 + 49 + 4 + 16 + 25 = 104$$

b. $\sum x = 22$, as found in Illustration A-2. Thus,

$$\left(\sum x\right)^2 = (22)^2 = 484$$

As you can see, there is quite a difference between $\sum x^2$ and $\left(\sum x\right)^2$. ∆∆

Likewise, $\sum xy$ and $\sum x \sum y$ are different. These forms will appear only when there are paired data, as shown in the following illustration.

∇ **ILLUSTRATION A-4**

Given the five pairs of data shown in the following table, find (a) $\sum xy$ and (b) $\sum x \sum y$.

x	1	6	9	3	4
y	7	8	2	5	10

Solution

a. $\sum xy$ means to sum the products of the corresponding x and y values. Therefore, we have

x	1	6	9	3	4
y	7	8	2	5	10
xy	7	48	18	15	40

$$\sum xy = 7 + 48 + 18 + 15 + 40 = 128$$

b. $\Sigma x \Sigma y$ means the product of the two summations, Σx and Σy. Therefore, we have

$$\sum x = 1 + 6 + 9 + 3 + 4 = 23$$
$$\sum y = 7 + 8 + 2 + 5 + 10 = 32$$
$$\sum x \sum y = (23)(32) = 736 \qquad \Delta\Delta$$

There are three basic rules for algebraic manipulation of the Σ notation.

NOTE c represents any constant value.

$$\boxed{\textbf{RULE 1: } \sum_{i=1}^{n} c = nc}$$

To prove this rule, we need only write down the meaning of $\sum_{i=1}^{n} c$:

$$\sum_{i=1}^{n} c = \underbrace{c + c + c + \ldots + c}_{n \text{ addends}}$$

Therefore,

$$\sum_{i=1}^{n} c = n \cdot c$$

∇ ILLUSTRATION A-5

Show that $\displaystyle\sum_{i=1}^{5} 4 = (5)(4) = 20$.

Solution

$$\sum_{i=1}^{5} 4 = \underbrace{4_{(\text{when } i=1)} + 4_{(\text{when } i=2)} + 4_{(i=3)} + 4_{(i=4)} + 4_{(i=5)}}_{\text{five 4s added together}}$$

$$= (5)(4) = 20 \qquad \Delta\Delta$$

$$\boxed{\text{RULE 2:} \qquad \sum_{i=1}^{n} cx_i = c \cdot \sum_{i=1}^{n} x_i}$$

To demonstrate the truth of Rule 2, we will need to expand the term $\sum_{i=1}^{n} cx_i$, and then factor our the common term c.

$$\sum_{i=1}^{n} cx_i = cx_1 + cx_2 + cx_3 + \cdots + cx_n$$
$$= c\left(x_1 + x_2 + x_3 + \cdots + x_n\right)$$

Therefore,

$$\sum_{i=1}^{n} cx_i = c \cdot \sum_{i=1}^{n} x_i$$

$$\boxed{\text{RULE 3:} \qquad \sum_{i=1}^{n} \left(x_i + y_i\right) = \sum_{i=1}^{n} x_i + \sum_{i=1}^{n} y_i}$$

The expansion and regrouping of $\sum_{i=1}^{n} \left(x_i + y_i\right)$ is all that is needed to show this rule.

$$\sum_{i=1}^{n} \left(x_i + y_i\right) = \left(x_1 + y_1\right) + \left(x_2 + y_2\right) + \cdots + \left(x_n + y_n\right)$$
$$= \left(x_1 + x_2 + \cdots + x_n\right) + \left(y_1 + y_2 + \cdots + y_n\right)$$

Therefore,

$$\sum_{i=1}^{n} \left(x_i + y_i\right) = \sum_{i=1}^{n} x_i + \sum_{i=1}^{n} y_i$$

▽ ILLUSTRATION A-6

Show that $\sum_{i=1}^{3} \left(2x_i + 6\right) = 2 \cdot \sum_{i=1}^{3} x_i + 18$.

Solution

$$\sum_{i=1}^{3} \left(2x_i + 6\right) = \left(2x_1 + 6\right) + \left(2x_2 + 6\right) + \left(2x_3 + 6\right)$$

$$= \left(2x_1 + 2x_2 + 2x_3\right) + (6 + 6 + 6)$$

$$= (2)\left(x_1 + x_2 + x_3\right) + (3)(6)$$

$$= 2\sum_{i=1}^{3} x_i + 18 \qquad \Delta\Delta$$

▽ **ILLUSTRATION A-7**

Let $x_1 = 2$, $x_2 = 4$, $x_3 = 6$, $f_1 = 3$, $f_2 = 4$, and $f_3 = 2$.
Find $\displaystyle\sum_{i=1}^{3} x_i \cdot \sum_{i=1}^{3} f_i$.

Solution

$$\sum_{i=1}^{3} x_i \cdot \sum_{i=1}^{3} f_i = \left(x_1 + x_2 + x_3\right) \cdot \left(f_1 + f_2 + f_3\right)$$

$$= (2 + 4 + 6) \cdot (3 + 4 + 2)$$

$$= (12)(9) = 108 \qquad \Delta\Delta$$

▽ **ILLUSTRATION A-8**

Using the same values for the x's and f's as in Illustration
A-7, find $\sum(xf)$.

Solution Recall that the use of no index numbers means "use all data."

$$\sum (xf) = \sum_{i=1}^{3} \left(x_i f_i\right) = \left(x_1 f_1\right) + \left(x_2 f_2\right) + \left(x_3 f_3\right)$$

$$= (2 \cdot 3) + (4 \cdot 4) + (6 \cdot 2) = 6 + 16 + 12 = 34$$

$$\Delta\Delta$$

∇∆ EXERCISES

A.1 Write each of the following in expanded form (without the summation sign):

a. $\displaystyle\sum_{i=1}^{4} x_i$ b. $\displaystyle\sum_{i=1}^{3} \left(x_i\right)^2$ c. $\displaystyle\sum_{i=1}^{5} \left(x_i + y_i\right)$

d. $\displaystyle\sum_{i=1}^{5} \left(x_i + 4\right)$ e. $\displaystyle\sum_{i=1}^{8} x_i y_i$ f. $\displaystyle\sum_{i=1}^{4} x_i^2 f_i$

A.2 Write each of the following expressions as summations, showing the subscripts and the limits of summation:

a. $x_1 + x_2 + x_3 + x_4 + x_5 + x_6$

b. $x_1 y_1 + x_2 y_2 + x_3 y_3 + \cdots x_7 y_7$

c. $x_1^2 + x_2^2 + \cdots x_9^2$

d. $\left(x_1 - 3\right) + \left(x_2 - 3\right) \cdots + \left(x_n - 3\right)$

A.3 Show each of the following to be true:

a. $\displaystyle\sum_{i=1}^{4} \left(5x_i + 6\right) = 5 \cdot \sum_{i=1}^{4} x_i + 24$

b. $\displaystyle\sum_{i=1}^{n} \left(x_i - y_i\right) = \sum_{i=1}^{n} x_i - \sum_{i=1}^{n} y_i$

A.4 Given $x_1 = 2$, $x_2 = 7$, $x_3 = -3$, $x_4 = 2$, $x_5 = -1$, and $x_6 = 1$, find each of the following:

a. $\displaystyle\sum_{i=1}^{6} x_i$ b. $\displaystyle\sum_{i=1}^{6} x_i^2$ c. $\left(\displaystyle\sum_{i=1}^{6} x_i\right)^2$

A.5 Given $x_1 = 4$, $x_2 = -1$, $x_3 = 5$, $f_1 = 4$, $f_2 = 6$, $f_3 = 2$, $y_1 = -3$, $y_2 = 5$, and $y_3 = 2$, find each of the following:

a. $\sum x$ b. $\sum y$ c. $\sum f$ d. $\sum (x - y)$

e. $\sum x^2$ f. $\left(\sum x\right)^2$ g. $\sum xy$ h. $\sum x \cdot \sum y$

i. $\sum xf$ j. $\sum x^2 f$ k. $\left(\sum xf\right)^2$

A.6   Suppose that you take out a $12,000 small-business loan.
       The terms of the loan are that each month for 10 years
       (120 months) you will pay back $100 plus accrued
       interest. The accrued interest is calculated by
       multiplying 0.005 (6 percent/12) times the amount of the
       loan still outstanding. That is, the first month you pay
       $12,000 \times 0.005$ in accrued interest, the second month
       $(\$12,000 - 100) \times 0.005$ in interest, the third month
       $[\$12,000 - (2)(100)] \times 0.005$, and so forth. Express the total
       amount of interest paid over the life of the loan by
       using summation notation.

The answers to these exercises can be found in the back of the
manual.

USING THE RANDOM NUMBER TABLE

The random number table is a collection of random digits. The term *random* means that each of the 10 digits (0,1,2,3,...,9) has an equal chance of occurrence. The digits, in Table 1 (Appendix B, JES2-p515) can be thought of as single-digit numbers (0-9), as two-digit numbers (00-99), as three-digit numbers (000-999), or as numbers of any desired size. The digits presented in Table 1 are arranged in pairs and grouped into blocks of five rows and five columns. This format is used for convenience. Tables in other books may be arranged differently.

Random numbers are used primarily for one of two reasons: (1) to identify the source element of a population (the source of data) or (2) to simulate an experiment.

∇ ILLUSTRATION 1

A simple random sample of 10 people is to be drawn from a population of 7564 people. Each person will be assigned a number, using the numbers from 0001 to 7564. We will view Table 1 as a collection of four-digit numbers (two columns used together), where the numbers 0001, 0002, 0003, ..., 7564 identify the 7564 people. The numbers 0000, 7565, 7566, ..., 9999 represent no one in our population; that is, they will be discarded if selected.

Now we are ready to select our 10 people. Turn to Table 1 (Appendix B, JES2-p515). We need to select a starting point and a path to be followed. Perhaps the most common way to locate a starting point is to look away and arbitrarily point to a starting point. The number we located this way was 3909 on page 516. (It is located in the upper left corner of the block that is in the fourth large block from the left and the second large block down.) From here we will proceed down the column, then go to the top of the next set of columns, if necessary. The person identified by number 3909 is the first source of data selected. Proceeding down the column, we find 8869 next. This number is discarded. The number 2501 is next. Therefore, the person identified by 2501 is the second source of data to be selected. Continuing down this column, our sample will be obtained from those people identified by the numbers 3909, 2501, 7485, 0545, 5252, 5612, 0997, 3230, 1051, 2712. (The numbers 8869, 8338, and 9187 were discarded.) ΔΔ

▽ ILLUSTRATION 2

Let's use the random number table and simulate 100 tosses of a coin. The simulation is accomplished by assigning numbers to each of the possible outcomes of a particular experiment. The assignment must be done in such a way as to preserve the probabilities. Perhaps the simplest way to make the assignment for the coin toss is to let the even digits (0,2,4,6,8) represent heads and the odd digits (1,3,5,7,9) represent tails. The correct probabilities are maintained:

$$P(H) = P(0,2,4,6,8) = \frac{5}{10} = 0.5 \text{ and } P(T) = P(1,3,5,7,9) = \frac{5}{10} = 0.5 .$$

Once this assignment is complete, we are ready to obtain our sample.

Since the question asked for 100 tosses and there are 50 digits to a "block" in Table 1 (Appendix B, JES2-p515), let's select two blocks as our sample of random one-digit numbers (instead of a column 100 lines long). Let's look away and point to one block on p. 516 and then do the same to select one block from p. 517. We picked the second block down in the first column of blocks on p. 516 (24 even and 26 odd numbers) and the second block down in the third column of blocks on p. 517 (23 even and 27 odd numbers). Thus we obtain a sample of 47 heads and 53 tails for our 100 simulated tosses.   ΔΔ

There are, of course, many ways to use the random number table. You must use your good sense in assigning the numbers to be used and in choosing the path to be followed through the table. One bit of advice is to make the assignments in as simple and easy a method as possible to avoid errors.

1. A random sample of size 8 is to be selected from a population that contains 75 elements. Describe how the random sample of the 8 objects could be made with the aid of the random number table.

2. A coin-tossing experiment is to be simulated. Two coins are to be tossed simultaneously and the number of heads appearing is to be recorded for each toss. Ten such tosses are to be observed. Describe two ways to use the random number table to simulate this experiment.

3. Simulate five rolls of three dice by using the random number table.

The answers to these exercises can be found in the back of the manual.

ROUND-OFF PROCEDURE

When rounding off a number, we use the following procedure.

STEP 1   Identify the position where the round-off is to occur. This is shown by using a vertical line that separates the part of the number to be kept from the part to be discarded. For example,

125.267 to the nearest tenth is written as   25.2|67

7.8890   to the nearest hundredth is written as 7.88|90

STEP 2   Step 1 has separated all numbers into one of four cases. (*X*'s will be used as placeholders for number values in front of the vertical line. These *X*'s can represent any number value.)

Case I: *XXXX*|000...

Case II: *XXXX*|--- (any value from 000...1 to 499...9)

Case III: *XXXX*|5000...0

Case IV: *XXXX*|--- (any value from 5000...1 to 999...9)

STEP 3   Perform the rounding off.

Case I requires no round-off. It's exactly *XXXX*.

▽ ILLUSTRATION 1

Round 3.5000 to the nearest tenth.

3.5|000 becomes 3.5

ΔΔ

Case II requires rounding. We will round down for this case. That is, just drop the part of the number that is behind the vertical line.

ILLUSTRATION 2

Round 37.6124 to the nearest hundredth.

37.61|24 becomes 37.61   ΔΔ

Case III requires rounding. This is the case that requires special attention. **When a 5 (exactly a 5) is to be rounded off, round to the even digit.** In the long run, half of the time the 5 will be preceded by an even digit $(0,2,4,6,8)$ and you will round down, while the other half of the time the 5 will be preceded by an odd digit $(1,3,5,7,9)$ and you will round up.

▽ ILLUSTRATION 3

Round 87.35 to the nearest tenth.

87. 3|5   becomes   87.4

Round 93.445 to the nearest hundredth.

3. 44|5 becomes 93.44

(**Note:** 87.35 is 87.35000... and 93.445 is 93.445000...)   ∆∆

Case IV requires rounding. We will round up for this case. That is, we will drop the part of the number that is behind the vertical line and we will increase the last digit in front of the vertical line by one.

▽ ILLUSTRATION 4

Round 7.889 to the nearest tenth.

7. 8|89   becomes   7.9   ∆∆

NOTE   **Case I, II, and IV describe what is commonly done. Our guidelines for Case III are the only ones that are different from typical procedure.**

If the typical round-off rule $(0, 1, 2, 3, 4$ are dropped; $5, 6, 7, 8, 9$ are rounded up) is followed, then $(n + 1) / (2n + 1)$ of the situations are rounded up. (n is the number of different sequences of digits that fall into each of Case II and Case IV.) That is more than half. You (as many others have) may say, "So what?" In today's world that tiny, seemingly insignificant amount becomes very significant when applied repeatedly to large numbers.

1.   Round each of the following to the nearest integer:
      a. 12.94   b. 8.762   c. 9.05   d. 156.49
      e. 45.5   f. 42.5   g. 102.51   h. 16.5001

2.   Round each of the following to the nearest tenth:
      a. 8.67   b. 42.333   c. 49.666   d. 10.25
      e. 10.35   f. 8.4501   g. 27.35001   h. 5.65

      i. 3.05   j. $\frac{1}{4}$

3.   Round each of the following to the nearest hundredth:
      a. 17.6666   b. 4.444   c. 54.5454   d. 102.055
      e. 93.225   f. 18.005   g. 18.015   h. 5.555

      i. 44.7450   j. $\frac{2}{3}$

The answers to these exercises can be found in the back of the manual.

REVIEW LESSONS

THE COORDINATE-AXIS SYSTEM AND THE EQUATION OF A STRAIGHT LINE

The rectangular coordinate-axis system is a graphic representation of points. Each point represents an ordered pair of values. (Ordered means that when values are paired, one value is always listed first, the other second.) The pair of values represents a horizontal location (the *x*-value, called the abscissa) and a vertical location (the *y*-value, the ordinate) in a fixed reference system. This reference system is a pair of perpendicular real number lines whose point of intersection is the 0 of each line (Figure 1-1, below).

Any point (x, y) is located by finding the point that satisfies both positional values. For example, the point P(2,3) is exactly 2 units to the right of 0 along the horizontal axis and 3 units above 0 along the vertical axis. Figure 1-2 (next page) shows two lines, *A* and *B*. Line *A* represents all the points that are 2 units to the right of 0 along the *x*-(horizontal) axis. Line *B* represents all the points that are 3 units above the 0 along the *y*-(vertical) axis. Point *P* is the one point that satisfies both conditions. Typically we think of locating point *P* by moving along the *x*-axis 2 units in the positive direction and then moving parallel to the *y*-axis 3 units in the positive direction (Figure 1-3, next page).

If either value is negative, we just move in a negative direction a distance equal to the number value.

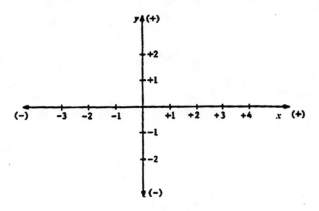

Figure 1-1

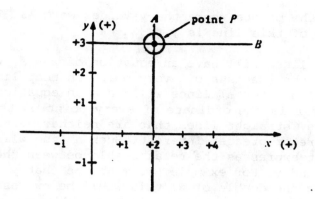

Figure 1-2

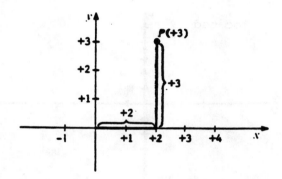

Figure 1-3

∇Δ EXERCISES

1. On a rectangular coordinate axis, drawn on graph paper,
   locate the following points.

   A(5,2)   B(−5,2)   C(−3,−2)

   D(3,0)   E(0,−2)   F(−2,5)

The equation of a line on a coordinate-axis system is a
statement of fact about the coordinates of all points that lie
on that line. This statement may be about one of the variables
or about the relationship between the two variables. In Figure
1-2 above, a vertical line was drawn at x = +2. A statement
that could be made about this line is that every point on it
has an x-value of 2. Thus, the equation of this line is x = 2.
The horizontal line that was drawn on the same graph passed

through all the points where the *y*-values were +3. Therefore the equation of this line is $y = +3$.

All vertical lines will have an equation of $x = a$, where *a* is the value of the abscissa of every point on that line. Likewise, all horizontal lines will have an equation of $y = b$, where *b* is the ordinate of every point on that line. In statistics, straight lines that are neither vertical nor horizontal are of greater interest. Such a line will have an equation that expresses the relationship between the two variables *x* and *y*. For example, it might be that *y* is always one less than the double of *x*; this would be expressed equationally as $y = 2x - 1$. There are an unlimited number of ordered pairs that fit this relationship; to name a few: $(0,-1)$, $(1,1)$, $(2,3)$, $(-2,-5)$, $(1.5,2)$, $(2.13,3.26)$ (see Figure 1-4).

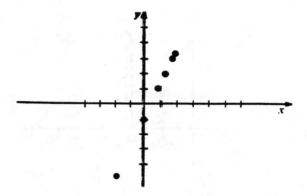

Figure 1-4

Notice that these points fall on a straight line. Many more points could be named that also fall on this same straight line; in fact, all the pairs of values that satisfy $y = 2x - 1$ lie somewhere along it. The converse is also true: all the points that lie on this straight line have coordinates that make the equation $y = 2x - 1$ a true statement. When drawing the line that represents the equational relationship between the coordinates of points, one needs to find only two points of the straight line; however, it is often useful to locate three of four to ensure accuracy.

EXERCISES (CONTINUED)

2. a. Find the missing values in the accompanying chart of ordered pairs, where $x + y = 5$ is the relationship.

x	y
-2	
0	
	4
3.5	
	-1

   b. On a coordinate axis, locate the same five points and then draw a straight line that passes through all of them.

3. Find five points that belong to the relationship expressed by $y = (3x/2) - 4$; then locate them on the axis system and draw the line representing $y = (3x/2) - 4$.

The form of the equation of a straight line that we are interested in is called the slope-intercept form. Typically in mathematics this slope-intercept form is expressed by $y = mx + b$, where m represents the concept of *slope* and b is the *y-intercept*.

Let's look at the y-intercept first. If you will look back at Exercise 3, you will see that the y-intercept for $y = (3x/2) - 4 \left[y = (3x/2) + (-4) \right]$ is -4. This value is simply the value of y at the point where the graph of the line intersects the y-axis, and all nonvertical lines will have this property. The value of the y-intercept may be found on the graph or from the equation. From the graph it is as simple as identifying it, but we will need to have the equation solved for y in order to identify it from the equation. For example, in Exercise 2 we had the equation $x + y = 5$; if we solve for y, we have $y = -x + 5$, and the y-intercept is 5 (the same as is found on the graph for Exercise 2).

The *slope* of a straight line is a measure of its inclination. This measure of inclination can be defined as the amount of vertical change that takes place as the value of *x* increases by exactly one unit. This amount of change may be found anywhere on the line since this value is the same everywhere on a given straight line. If we inspect the graph drawn for Exercise 2 (Figure 1-5), we will see that the slope is −1, meaning that each *x* increase of 1 unit results in a decrease of 1 unit in the *y*-value.

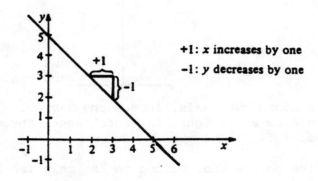

+1: *x* increases by one

−1: *y* decreases by one

Figure 1-5

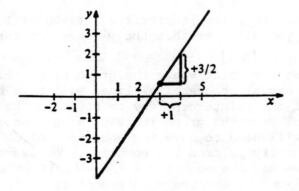

Figure 1-6

An inspection of the graph drawn for Exercise 3 (Figure 1-6) will reveal a slope of +3/2. This means that y increased by 3/2 for every increase of one unit in x.

Illustration: Find the slope of the straight line that passes through the points $(-1,1)$ and $(4,11)$.

Solution: $m = \dfrac{\Delta y}{\Delta x} = \dfrac{11 - 1}{4 - (-1)} = \dfrac{10}{5} = 2$

NOTES: 1. $\Delta y = y_2 - y_1$ and $\Delta x = x_2 - x_1$ where (x_1, y_1) and (x_2, y_2) are the two points that the line passes through.

2. As stated before, these properties are both algebraic and graphic. If you know about them from one source, then the other must agree. Thus, if you have the graph, you should be able to read these values from it, or if you have the equation, you should be able to draw the graph of the line with these given properties.

ILLUSTRATION: Graph $y = 2x + 1$.

SOLUTION: $m = 2$ and $b = 1$. Locate the y-intercept $(y = +1)$ on the y-axis. Then draw a line that has a slope of 2 (Figure 1-7).

Figure 1-7

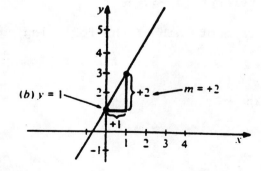

Illustration: Find the equation of the line that passes through $(-1,1)$ and $(2,7)$.

Solution: (Algebraically) $m = \dfrac{\Delta y}{\Delta x} = \dfrac{7-1}{2-(-1)} = \dfrac{6}{3} = 2$

$y = 2x + b$, and the line passes through $(2,7)$. Therefore $x = 2$ and $y = 7$ must satisfy (make the statement true) $y = 2x + b$. In order for that to happen, b must be equal to $3[7 = 2(2) + b]$. Therefore the equation of such a line is $y = 2x + 3$.

(Graphically) Draw a graph of a straight line that passes through $(-1,1)$ and $(2,7)$; then read m and b from it (see Figure 1-8).

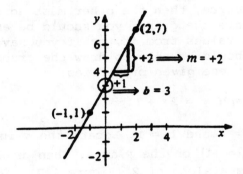

Figure 1-8

Therefore $y = mx + b$ becomes $y = 2x + 3$.

EXERCISES (Continued)

4. Write the equation of a straight line whose slope is 10 and whose y-intercept is -3.

5. Draw the graph of each of the following equations (use graph paper).
   a. $y = x + 2$
   b. $y = -2x + 10$
   c. $y = (1/3)x - 2$

6. Find the equation of the straight line that passes through each of the following pairs of points (a) algebraically and (b) graphically (use graph paper).

   I.   (3,1) and (9,5)

   II.   (−2,3) and (6,−1)

The preceding discussion about the equation of a straight line is presented from a mathematical point of view. Mathematicians and statisticians often approach concepts differently. For instance, the statistician typically places the terms of a linear equation in exactly the opposite order from the mathematician's equation.

To the statistician, for example, the equation of the straight line is $y = b + mx$, while to the mathematician it is $y = mx + b$. m and b represent exactly the same concepts in each case — the different order is a matter of emphasis. The mathematician's first interest in an equation is usually the highest-powered term; thus he or she places it first in the sequence. The statistician tends to describe a relationship in as simple a form as possible; thus his or her first interest is usually the lower-powered terms. The equation of the straight line in statistics is $y = b_0 + b_1 x$, where b_0 is the y-intercept and b_1 represents the slope.

The answers to the exercises can be found in the back of the manual.

TREE DIAGRAMS

The purpose of this lesson is to learn how to construct and read a tree diagram. A tree diagram is a drawing that schematically represents the various possible outcomes of an experiment. It is called a tree diagram because of the branch concept that it demonstrates.

Let's consider the experiment of tossing one coin one time. We will start the experiment by tossing the coin and will finish it by observing a result (heads or tails)(see Figure 2-1).

Start   **Observation**
                    Heads
                    Tails

Figure 2-1

This information is expressed by the tree shown in Figure 2-2.

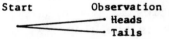

Figure 2-2

In reading a diagram like this, the single point at the left is simply interpreted as we are ready to start and do not yet know the outcome. The branches starting from this point must represent all of the different possibilities. With one coin there are only two possible outcomes, thus two branches.

∇∆ **EXERCISES**

1. Draw a tree diagram that shows the possible results from rolling a single die once.

2. Draw a tree diagram showing the possible methods of transportation that could be used to travel to a resort area. The possible choices are car, bus, train, and airplane.

Now let's consider the experiment of tossing a coin and single
die at the same. What are the various possibilities? The coin
can result in a head (H) or a tail (T) and the die could show
a 1, 2, 3, 4, 5, or 6. Thus to show all the possible pairs of
outcomes, we must decide which to observe first. This is an
arbitrary decision, as the order observation does not affect
the possible pairs of results. To construct the tree to
represent this experiment we list the above-mentioned
possibilities in columns, as shown in Figure 2-3.

Start	Coin	Die
		1
		2
	H	3
		4
		5
		6
	T	

Figure 2-3

From "start" we draw two line segments that represent the
possibility of *H* or *T* (Figure 2-4). The top branch means that
we might observe a head on the coin. Paired with it is the
result of the die, which could be any of the six numbers. We
see this represented in Figure 2-5.

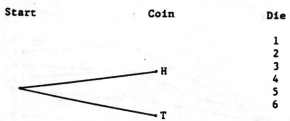

Figure 2-4

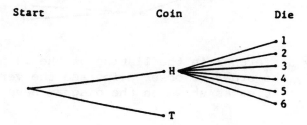

Figure 2-5

However, the outcome could have been *T*, so *T* must have branches going to each of the numbers 1 through 6. To make the diagram easier to read we list these outcomes again and draw another set of branches (Figure 2-6).

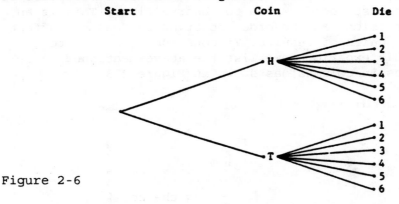

Figure 2-6

Figure 2-6 shows 12 different possible pairs of results — each one of the 12 branches on the tree represents one of these pairs. (A complete branch is a path from the start to an end.) The 12 branches in Figure 2-6, from top to bottom, are H1, H2, H3, H4, H5, H6, T1, T2, T3, T4, T5, T6.

The ordering could have been reversed: the tree diagram in Figure 2-7 shows the die result first and the coin result second.

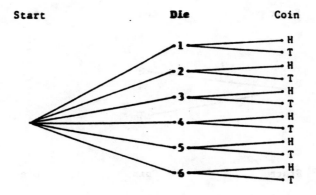

Figure 2-7

Is there any difference in the listing of the 12 possible results? Only the order of observation and the vertical order of the possibilities as shown on the diagram have changed.

-- 318 --

They are still the same 12 pairs of possibilities. If the experiment contains more than two stages of possible events we may expand this tree as far as needed.

EXERCISES (CONTINUED)

3. a. Draw a tree diagram that represents the possible results from tossing two coins.
   b. Repeat part (a) considering one coin a nickel and the other a penny.
   c. Is the list of possibilities for part (b) any different than it was for part (a)?

4. Draw a tree diagram representing the tossing of the coins.

5. a. Draw a tree diagram representing the possible results that could be obtained when two dice are rolled.
   b. How many branch ends does your tree have?

On occasion the stages of an experiment will be ordered; when this is the case the tree diagram must show ordered sets of branches, as in Figure 2-8, next page. The experiment consists of rolling a die. Then the result of the die will dictate your next trial. If an odd number results, you will toss coin. If a two or a six occurs, you stop. If any other number (a four) occurs, you roll the die again.
Notice that the tree diagram becomes a very convenient "road map" showing all the various possibilities that may occur in an experiment of this nature. Remember that an event is represented by a complete branch (a broken line from the start to an end), and the number of ends of branches is the same as the number of possibilities for the experiment. There are 14 branches in the tree diagram in Figure 2-8. Do you agree?

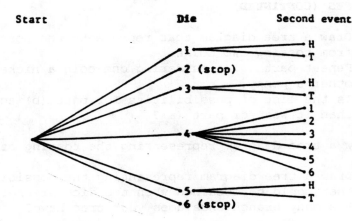

Start	Die	Second event

Figure 2-8

EXERCISES (CONTINUED)

6. Students at our college are to be classified as male or female, graduates of public or private high schools, and by the type of curriculum they are enrolled in, liberal arts or career. Draw a tree diagram which shows all of the various possible classifications.

7. There are two scenic routes (A and B) as well as one business route (C) by which you may travel from your home to a nearby city. You are planning to drive to that city by way of one route and come home by a different route.
   a. Draw a tree diagram representing all of your possible choices for going and returning.
   b. How many different trips could you plan?
   c. How many of these trips are scenic in both directions?

The answers to the exercises can be found in the back of the manual.

VENN DIAGRAMS

The Venn diagram is a useful tool for representing sets. It is a pictorial representation that uses geometric configurations to represent *set containers*. For example, a set might be represented by a circle — the circle acts like a "fence" and encloses all of the elements that belong to that particular set. The figure drawn to represent a set must be closed, and the elements are either inside the boundary and belong to that set or they are outside and do not belong to that set. The universal set (sample space or population) is generally represented by a rectangular area, and its subsets are generally circles inside the rectangle. Complements, intersections, and unions of sets then become regions of various shapes as prescribed by the situation. The Venn diagram in Figure 3-1 shows a universal set and a subset P.

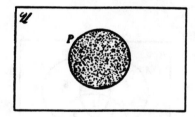

Figure 3-1

Any element that is represented by a point inside the rectangle is an element of the universal set. Likewise, any element represented by a point inside the circle, P(the shaded area), is a member of set P. The unshaded area of the rectangle then represents $\overline{P}$.

The Venn diagrams in Figure 3-2 show the regions representing $A \cap B$, $A \cup B$, $\left(\overline{A \cap B}\right)$, and $\left(\overline{A \cup B}\right)$. The shaded regions represent the identified sets.

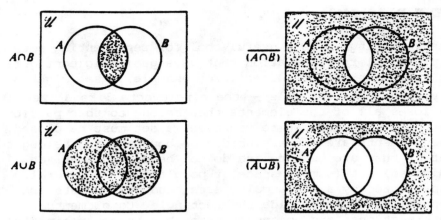

Figure 3-2

When three subsets of the same population are being discussed, three circles can be used to represent all of the various possible situations.

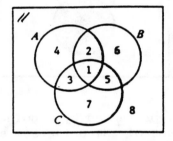

Figure 3-3

In Figure 3-3 the eight regions that are formed by intersecting the three sets have been numbered for convenience. Each of these regions represents the intersection of three sets (sets and/or complements of sets), as shown below.

Region Number	Set Representation
1	$A \cap B \cap C$
2	$A \cap B \cap \overline{C}$
3	$A \cap \overline{B} \cap C$
4	$A \cap \overline{B} \cap \overline{C}$
5	$\overline{A} \cap B \cap C$
6	$\overline{A} \cap B \cap \overline{C}$
7	$\overline{A} \cap \overline{B} \cap C$
8	$\overline{A} \cap \overline{B} \cap \overline{C}$

Region 1 $(A \cap B \cap C)$ might be thought of as the set of elements that belong to A, B, and C. Region 4 represents the set of elements that belong to A, $\overline{B}$ (but not to set B), and $\overline{C}$ (but not to set C). Region 8 represents the set of elements that belong to $\overline{A}$, $\overline{B}$, and $\overline{C}$ (or that do not belong to A, B, or C). The others can be described in similar fashion.

Figure 3-4 shows the union as sets B and C in the shaded areas of all three sets. Notice that $B \cup C$ is composed of regions 1, 2, 3, 5, 6, and 7. (You might note that three of these regions are inside A and three are outside).

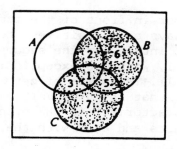

Figure 3-4

▽△ EXERCISES

1. Shade the regions that represent each of the following sets on a Venn diagram as shown in Figure 3-5.
   a. A
   b. B
   c. $A \cap B$
   d. $A \cup B$
   e. $\overline{A} \cup B$
   f. $\overline{A} \cup \overline{B}$

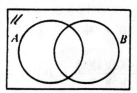

Figure 3-5

2. On a diagram showing three sets, P, Q, and R, shade the regions that represent the following sets.
   a. P
   b. $P \cap Q$
   c. $P \cup R$
   d. $\overline{P}$
   e. $P \cap Q \cap R$
   f. $P \cup \overline{Q}$
   g. $P \cup Q \cup R$
   h. $P \cup Q \cup \overline{R}$

The answers to these exercises can be found in the back of the manual.

THE USE OF FACTORIAL NOTATION

The factorial notation is a shorthand way to identify the product of a particular set of integers. 5! (five factorial) stands for the product of all positive integers starting with the integer 5 and proceeding downward (in value) until the integer 1 is reached. That is, $5! = 5 \times 4 \times 3 \times 2 \times 1$, which is 120. Likewise, $n!$ symbolizes the product of the integer n multiplied by the next smaller integer $(n - 1)$ multiplied by the next smaller integer $(n - 2)$ and so on, until the last integer, the number 1, is reached.

NOTES:
1. The number in front of the factorial symbol (!) will always be a positive integer or 0.
2. The last integer in the sequence is always the integer 1, with one exception: 0! (zero factorial). The value of zero factorial is defined to be 1, that is, $0! = 1$.
   1! (one factorial) is the product of a sequence that starts and ends with the integer 1, thus $1! = 1$.
   2! (two factorial) is the product of 2 and 1. That is, $2! = (2)(1) = 2$.
   $3! = (3)(2)(1) = 6$
   $n! = (n)(n - 1)(n - 2)(n - 3)...(2)(1)$
   $(n - 2)! = (n - 2)(n - 3)(n - 4)...(2)(1)$

$$(4!)(6!) = (4 \cdot 3 \cdot 2 \cdot 1)(6 \cdot 5 \cdot 4 \cdot 3 \cdot 2 \cdot 1) = (24)(720) = 17280$$
$$4(6!) = (4)(6!) = 4(6 \cdot 5 \cdot 4 \cdot 3 \cdot 2 \cdot 1) = (4)(720) = 2880$$

$$\frac{6!}{4!} = \frac{(6)(5)(4)(3)(2)(1)}{(4)(3)(2)(1)} = (6)(5) = 30$$

∇∆ Exercises

Evaluate each of the following factorials.

1. 4!   2. 6!

3. 8!   4. (6!)(8!)

5. $\dfrac{8!}{6!}$ 6. $\dfrac{8!}{4!\,4!}$

7. $\dfrac{8!}{6!\,2!}$ 8. $2\dfrac{8!}{[5!]}$

The answers to these exercises can be found in the back of the manual.

ANSWERS TO INTRODUCTORY CONCEPTS
AND REVIEW LESSONS EXERCISES

Summation Notation Exercises

1. (a) $x_1 + x_2 + x_3 + x_4$
    (b) $x_1^2 + x_2^2 + x_3^2$
    (c) $(x_1 + y_1) + (x_2 + y_2) + (x_3 + y_3) + (x_4 + y_4) + (x_5 + y_5)$
    (d) $(x_1 + 4) + (x_2 + 4) + (x_3 + 4) + (x_4 + 4) + (x_5 + 4)$
    (e) $x_1y_1 + x_2y_2 + x_3y_3 + x_4y_4 + x_5y_5 + x_6y_6 + x_7y_7 + x_8y_8$
    (f) $x_1^2f_1 + x_2^2f_2 + x_3^2f_3 + x_4^2f_4$

2. (a) $\sum\limits_{i=1}^{6} x_i$ (b) $\sum\limits_{i=1}^{7} x_iy_i$ (c) $\sum\limits_{i=1}^{9} (x_i)^2$ (d) $\sum\limits_{i=1}^{n} (x_i - 3)$

4. (a) 8   (b) 68   (c) 64

5. (a) 8   (b) 4   (c) 12   (d) 4   (e) 42
    (f) 64   (g) -7   (h) 32   (i) 20   (j) 120
    (k) 400

6. $\sum\limits_{i=1}^{120} [0.005(12,000 - (i-1)100)]$

Using the Random Number Table Exercises

2. (a) Use a two-digit number to represent the results
        obtained. Let the first digit represent one of the
        coins and the second digit represent the other coin.
        Let an even digit indicate heads and an odd digit
        tails. Observe 10 two-digit numbers from the table.
        If a 16 is observed, it represents tails and heads on
        two coins. One head was therefore observed. The
        probabilities have been preserved.
    (b) A second way to simulate this experiment is to find
        the probabilities associated with the various
        possible results. The number of heads that can be
        seen on two coins is 0, 1, or 2. (HH,HT,TH,TT is the
        sample space.) P(no heads) = 1/4; P(one head) = 1/2;

P(two heads) = 1/4. Using two-digit numbers, let the numbers 00 to 24 stand for no head appeared, 25 to 74 stand for one head appeared, and 75 to 99 stand for two heads appeared. The probabilities have again been preserved. Observe 10 two- digit numbers.

Round-Off Procedure Exercises

1. (a) 13   (b) 9   (c) 9   (d) 156   (e) 46
    (f) 42   (g) 103   (h) 17
2. (a) 8.7   (b) 42.3   (c) 49.7   (d) 10.2   (e) 10.4
    (f) 8.5   (g) 27.4   (h) 5.6   (i) 3.0   (j) 0.2

3. (a) 17.67   (b) 4.44   (c) 54.55   (d) 102.06   (e) 93.22
    (f) 18.00   (g) 18.02   (h) 5.56   (i) 44.74   (j) 0.67

Review Lessons

The Coordinate-Axis System and the Equation of a Straight Line

1.

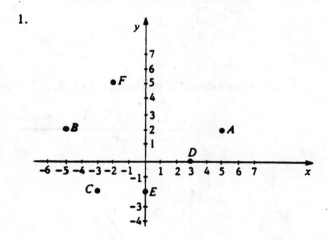

2. a.

Point	x	y
A	-2	7
B	0	5
C	1	4
D	3.5	1.5
E	6	-1

b.

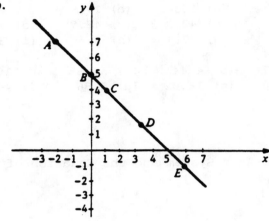

3. Pick any values of x you wish; x = -2, 0, 1, 2, 4 will be convenient.

x =	-2	0	1	2	4
y =	-7	-4	-2½	-1	2

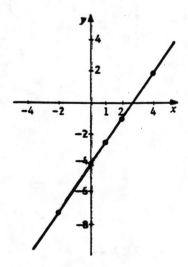

b.

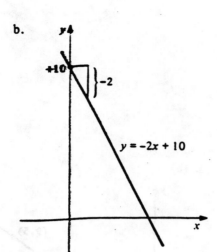

$y = -2x + 10$

c.

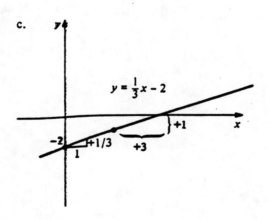

$y = \frac{1}{3}x - 2$

I. (b)

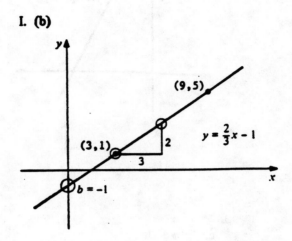

II. (a) Given points (-2,3) and (6,-1).

$$m = \frac{\Delta y}{\Delta x} = \frac{(-1) - (3)}{6 - (-2)} = \frac{-4}{8} = -\frac{1}{2}$$

$$y = \frac{1}{2}x + b \qquad [\text{use } (-2,3)]$$

$$3 = \left(-\frac{1}{2}\right)(-2) + b \text{ implies } b = 2$$

Thus $y = -\frac{1}{2}x + 2.$

(b)

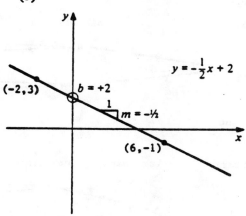

1.

Tree Diagrams

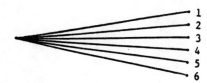

2.

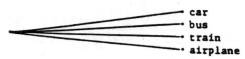

3. a.

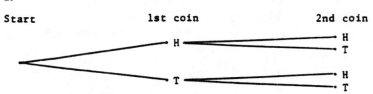

b.

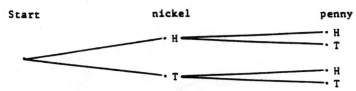

c. no

4.

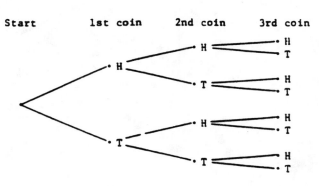

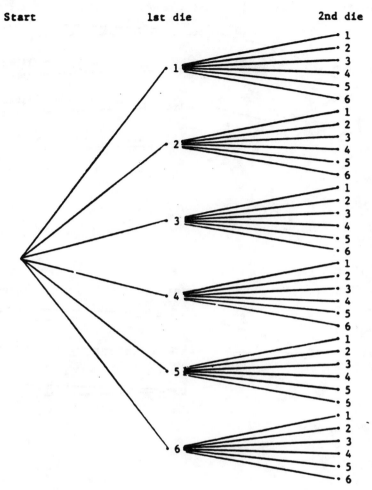

b. 36 branch ends

6.

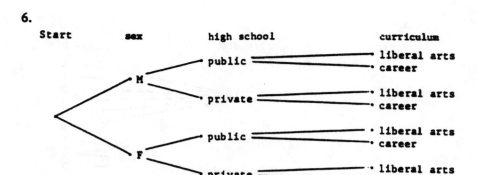

7. a.

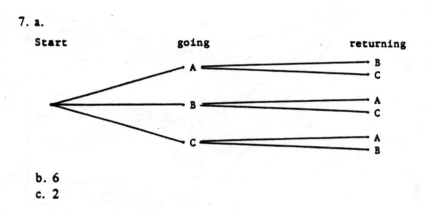

b. 6
c. 2

Venn Diagrams

1. a.   b.

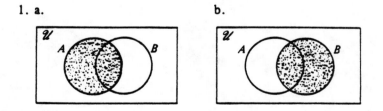

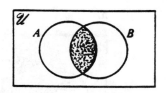

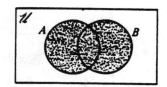

e.

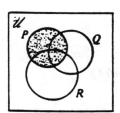

f.

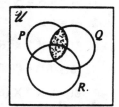

2. a.

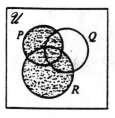

b.

c.

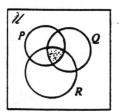

d.

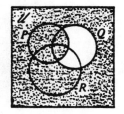

e.

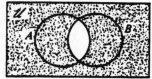

f.

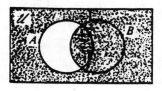

g.

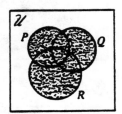

h.

The Use of Factorial Notation

1. $4! = 4 \times 3 \times 2 \times 1 = \underline{24}$

2. $6! = 6 \times 5 \times 4 \times 3 \times 2 \times 1 = \underline{720}$

3. $8! = 8 \times 7 \times (6!) = 56 \times 720 = \underline{40320}$

4. $(6!)(8!) = (720)(40320) = \underline{29,030,400}$

5. $\dfrac{8!}{6!} = \dfrac{8 \times 7 \times (6!)}{6!} = 8 \times 7 = \underline{56}$

6. $\dfrac{8!}{4!\,4!} = \dfrac{8 \times 7 \times 6 \times 5 \times (4!)}{4 \times 3 \times 2 \times 1 \times (4!)} = 2 \times 7 \times 5 = \underline{70}$

7. $\dfrac{8!}{6!\,2!} = \dfrac{8 \times 7 \times 6!}{(6!) \times 2 \times 1} = 4 \times 7 = \underline{28}$

8. $2\left(\dfrac{8!}{5!}\right) = 2\left(\dfrac{8 \times 7 \times 6 \times 5!}{5!}\right) = 2 \times 8 \times 7 \times 6 = \underline{672}$